NAGIOS

NAGIOS

Building Enterprise-Grade Monitoring Infrastructure for Systems and Networks

Second Edition

David Josephsen

PRENTICE
HALL

Upper Saddle River, NJ • Boston • Indianapolis • San Francisco
New York • Toronto • Montreal • London • Munich • Paris
Madrid • Cape Town • Sydney • Tokyo • Singapore • Mexico City

The publisher offers excellent discounts on this book when ordered in quantity for bulk purchases or special sales, which may include electronic versions and/or custom covers and content particular to your business, training goals, marketing focus, and branding interests. For more information, please contact:

U.S. Corporate and Government Sales
(800) 382-3419
corpsales@pearsontechgroup.com

For sales outside the United States, please contact:

International Sales
international@pearsoned.com

Visit us on the Web: informit.com/aw

Library of Congress Cataloging-in-Publication Data is on file.

ISBN-13: 978-0-13-313573-2
ISBN-10: 0-13-313573-X

Text printed in the United States on recycled paper at R.R. Donnelley in Crawfordsville, Indiana.

First Printing: April 2013

For Cynthia, for enduring and encouraging my incessant curiosity.
And for Tito, the cat with the biggest heart.

CONTENTS

FOREWORD

People often say that Nagios is "flexible," by which I think they mean that it is easily extended, but that misses the point. The power inherent in Nagios' design derives not from its extensibility, but rather from its insistence on being extended. This is an admittedly small but important distinction. Many pieces of software can be extended to do new things, but very few pieces of software do nothing until you've extended them, and it is exactly because of this—this inherent demand that you customize it to suit your needs—that Nagios has always been a synthesis of contributions from engineers and administrators working to solve their own individual problems. No two installations are alike, and that is by design.

In the years since I first created Nagios, it has grown in breadth and scope beyond anything I'd imagined. With over 1 million users worldwide, Nagios Core has found a home everywhere from huge Fortune-500 conglomerates to state of the art scientific research labs. The Nagios user community is one of the healthiest and most actively contributing open source communities out there, with nearly 4,000 published plug-ins, add-ons, and extensions—many of which are sufficiently complex to warrant books of their own. The community is so large, diverse, and active, that Nagios now has its own annual conference where contributors, users, and educators come together to share ideas, learn tips and tricks, and find out about upcoming developments in the project.

There is also a thriving community of corporations at work on extending and supporting Nagios. In 2007 I joined them, founding Nagios Enterprises. Our flagship product, Nagios XI, is both an evolutionary step forward, and (as it should be) a fully-reverse compatible extension to Nagios Core. XI embraces the extend-by-design lineage of Core, preserving the power and flexibility of Core, while expanding its accessibility and usability.

But even given the wonderful success Nagios has enjoyed, I'm the first to admit that flexibility comes with a price. It can be difficult for newcomers and experienced admins alike to build and deploy a successful monitoring solution, and many of the challenges have nothing whatsoever to do with computers. Luckily, David is one of the few technical writers that are able to cover a complex subject like this in an easy-to-understand format. Whether you're a newcomer to the world of network, system, and IT monitoring, or you're an experienced Nagios admin, David's work is sure to be helpful to you.

—**Ethan Galstad**, Nagios Founder and President

ACKNOWLEDGMENTS

My lovely wife, Cynthia, is patient and encouraging and pretty, and I love her.

Ethan Galstad, whose interest prompted the second edition, and without whom there would be no Nagios.

The tech reviewers on this project were outstanding—thanks, guys.

Last, my editors at Prentice Hall have been great. They aren't at all like the editors in Spiderman or Fletch. Debra Williams Cauley and Kim Boedigheimer are a hardworking, on the ball, and clued-in pair of professionals. They've been patient and helpful, and I appreciate their time and attention.

Thanks.

ABOUT THE AUTHOR

David Josephsen is the Director of Systems Engineering at DBG, Inc., where he maintains a collection of geographically dispersed server farms. He has more than a decade of hands-on experience with UNIX systems, routers, firewalls, and load balancers in support of complex, high-volume networks. In addition to this book, he authored several chapters in the O'Reilly book *Monitoring with Ganglia*, and currently writes "iVoyer," the systems monitoring column for *;login* magazine. Josephsen is just one of many thousands of avid Nagios users.

ABOUT THE TECHNICAL REVIEWERS

Mark Bainter

Mark Bainter leads a team of sysadmins providing outsourced monitoring and management of high volume mail systems for Message Systems' clients, leveraging over 15 years experience as a sysadmin specializing in systems integration, monitoring, and automation. He is an autodidactic polymath and impenitent sesquipedalian. Mark currently resides in Texas with his lovely wife and four children and in his free time he enjoys reading, woodworking, and losing at Settlers to his wife.

Mike Guthrie

Mike Guthrie is the lead developer at Nagios enterprises and has developed new features and add-ons for Nagios Core, Nagios XI, and Nagios Fusion. Mike does the bulk of his programming in PHP and particularly enjoys front-end web development and data visualizations. When he's not at work, he enjoys spending time with his family, being outside, and working on his house.

Mathias Kettner

Mathias Kettner is known as the author of Check_MK, MK Livestatus, and other Nagios add-ons. He runs a fast growing company in Munich, Germany, which is dedicated to system monitoring based on Nagios, and offers professional support and software development.

Introduction

This is a book about untrustworthy machines—machines, in fact, that are every bit as untrustworthy as they are critical to our well being. But I don't need to bore you with a laundry list of how prevalent computer systems have become or with horror stories about what can happen when they fail. If you picked up this book, I'm sure you're well aware of the problems: layer upon layer of interdependent libraries hiding bugs in their abstraction, script kiddies, viruses, DDOS attacks, hardware failure, end-user error, backhoes, hurricanes, and on and on. It doesn't matter whether the root cause is malicious or accidental; your systems will fail, and when they do, only two things will save you from the downtime: redundancy and monitoring systems.

Do It Right the First Time

In concept, monitoring systems are simple: an extra system or collection of systems whose job is to watch the other systems for problems. For example, the monitoring system could periodically connect to a Web server to make sure it responds and, if not, send notifications to the administrators. Although it sounds straightforward, monitoring systems have grown into expensive, complex pieces of software. Many now have agents larger than 500MB, include proprietary scripting languages, and sport price tags above $60,000.

When implemented correctly, a monitoring system can be your best friend. It can notify administrators of glitches before they become crises, help architects tease out patterns corresponding to chronic interoperability issues, and give engineers detailed capacity planning information. A good monitoring system will help the security guys correlate interesting events, show the network operations center personnel where the bandwidth bottlenecks are, and provide management with much needed high-level visibility into the critical systems that they bet their business on. A good monitoring system can help you uphold your service level agreement (SLA) and even take steps to solve problems without waking anyone up at all. Good monitoring systems save money, bring stability to complex environments, and make everyone happy.

When done poorly, however, the same system can wreak havoc. Bad monitoring systems cry wolf at all hours of the night so often that nobody pays attention anymore; they install backdoors into your otherwise secure infrastructure, leech time and resources away from other projects, and congest network links with megabyte upon megabyte of health checks. Bad monitoring systems can really suck.

Unfortunately, getting it right the first time isn't as easy as you might think, and in my experience, a bad monitoring system doesn't usually survive long enough to be fixed. Bad monitoring systems are too much of a burden on everyone involved, including the systems being monitored. In this context, it's easy to see why large corporations and governments employ full-time monitoring specialists and purchase software with six-figure price tags. They know how important it is to get it right the first time.

Small- to medium-sized businesses and universities can have environments as complex as or even more complex than large companies, but they obviously don't have the luxury of high-priced tools and specialized expertise. Getting a well-built monitoring infrastructure in these environments, with their geographically dispersed campuses and satellite offices, can be a challenge. But having spent a good part of the past 13 years building and maintaining monitoring systems, I'm here to tell you that not only is it possible to get it done right the first time, but you can do it for free, with a bit of elbow grease, some open source tools, and a pinch of imagination.

Why Nagios?

Nagios is, in my opinion, the best system and network monitoring tool available, open source or otherwise. Its modularity and straightforward approach to monitoring make it easy to work with and highly scalable. Further, Nagios's open source license makes it freely available and easy to extend to meet your specific needs. Instead of trying to do everything for you, Nagios excels at interoperability with other open source tools, which makes it very flexible. If you're looking for a monolithic piece of software with check boxes that solve all your problems, this probably isn't the book for you. But before you stop reading, give me another paragraph or two to convince you that the check boxes aren't really what you're looking for.

Most commercial offerings get it wrong because their approach to the problem assumes that everyone wants the same solution. To a certain extent, this is true. Everyone has a large glob of computers and network equipment and wants to be notified if some subset of it fails.

So if you want to sell monitoring software, the obvious way to go about it is to create a piece of software that knows how to monitor every conceivable piece of computer software and networking gear in existence. The more gadgets your system can monitor, the more people you can sell it to. To someone who wants to sell monitoring software, it's easy to believe that monitoring systems are turnkey solutions and whoever's software can monitor the largest number of gadgets wins.

The large commercial packages I've worked with all seem to follow this logic. Not unlike the Borg, they are methodically locating new computer gizmos and adding the requisite monitoring code to their solution—or worse, acquiring other companies who already know how to monitor lots of computer gadgetry and bolting those companies' code onto their own. They quickly become obsessed with features, creating enormous spreadsheets of supported gizmos. Their software engineers exist so that the presales engineers can come to your office and say to your managers, through seemingly layers of white gleaming teeth, "Yes, our software can monitor that."

The problem is that monitoring systems are not turnkey solutions. They require a large amount of customization before they start solving problems and herein lies the difference between people selling monitoring software and those designing and implementing monitoring systems. When you're trying to build a monitoring system, a piece of software that can monitor every gadget in the world by clicking a check box is not as useful to you as one that makes it easy to monitor what you need, in exactly the manner that you want. By focusing on *what* to monitor, the proprietary solutions neglect the *how*, which limits the context in which they may be used.

Take ping, for example. Every monitoring system I've ever dealt with uses ICMP Echo requests, otherwise known as pings, to check host availability in one way or another. But if you want to control *how* a proprietary monitoring system uses ping, architectural limitations become quickly apparent. Let's say I want to specify the number of ICMP packets to send, or I want to be able to send notifications based on the round-trip time of the packet in microseconds instead of simple pass/fail. More complex environments may necessitate that I use IPv6 pings, or that I portknock[1] before I ping. The problem with the monolithic, feature-full approach is that these changes represent changes to the core application logic and are, therefore, nontrivial to implement.

In the commercial monitoring applications I've worked with, if these ping examples could be performed at all, they would require reimplementing the ping logic in the monitoring system's proprietary scripting language. In other words, you would have to toss out the built-in ping functionality altogether. Perhaps being able to control the specifics of ping checks is of questionable value to you, but if you don't have any control over something as basic as ping, what are the odds that you'll have finite enough control over the most important checks in your environment? They've made the assumption that they know *how* you want to ping things and from then on it was game over; they never thought about it again. And why would they? The ping feature is already in the spreadsheet, after all.

When it comes to gizmos, Nagios's focus is on modularity. Single-purpose monitoring applets called plug-ins provide support for specific devices and services. Rather than participating in the feature arms race, hardware support is community driven. As community members have a need to monitor new devices or services, new plug-ins are written and usually more quickly than the commercial applications can add the same support. In practice, Nagios will always support everything you need it to and without ever needing to upgrade Nagios itself. Nagios also provides the best of both worlds when it comes to support, with several commercial options, as well as a thriving and helpful community that provides free support through various forums and mailing lists.

Choosing Nagios as your monitoring platform means that your monitoring effort will be limited by your own imagination, technical prowess, and political savvy. Nagios can go anywhere you want it to and the trip there is usually pretty simple. Although Nagios can do everything the commercial applications can, and more, without the bulky, insecure agent install, it usually doesn't compare favorably to commercial monitoring systems because when spreadsheets are parsed, Nagios doesn't have as many checks. In fact, if they're counting correctly, Nagios has no checks at all, because technically it doesn't know *how* to monitor anything; it prefers that you tell it how. The question of "how" is difficult to encompass with a check box.

What's in This Book?

Although Nagios is the biggest piece of the puzzle, it's only one of the myriad of tools that make up a world-class open source monitoring system. With several books, superb online documentation, and lively and informative mailing lists, it's also the best-documented piece of the puzzle. So my intention in writing this book is to pick up where the documentation leaves off. This is not a book about Nagios as much as it is a book about the construction of monitoring systems using Nagios, and there is much more to building monitoring systems than configuring a monitoring tool.

I'll cover the usual configuration boilerplate, but configuring and installing Nagios is not my primary focus. Instead, to help you build great monitoring systems, I need to introduce you to the protocols and tools that enhance Nagios's functionality and simplify its configuration. I need to give you an in-depth understanding of the inner workings of Nagios itself, so you can extend it to do whatever you might need. I need to spend some time in this book exploring possibilities because Nagios is limited only by what you feel it can do.

Finally, I need to write about things only loosely related to Nagios, like best practices, SNMP, visualizing time-series data, and various Microsoft scripting technologies, such as WMI and WSH.

Most important, I need to document Nagios itself in a different way. By introducing it in terms of a task-efficient scheduling and notification engine, I can keep things simple while talking about the internals up front. Rather than relegating important information to the seldom-read advanced section, I'll empower you early by covering topics like plug-in customization and scheduling as core concepts.

Although the chapters more or less stand on their own, and I've tried to make the book as reference-friendly as possible, I think it reads better as a progression from start to finish. I encourage you to read from cover to cover, skipping over anything you are already familiar with. The text is not large, but I think you'll find it dense with information and even the most seasoned monitoring veterans should find more than a few useful nuggets of wisdom.

The chapters tend to build on each other and casually introduce Nagios-specific details in the context of more general monitoring concepts. Because many important decisions need to be made before any software is installed, I begin with "Best Practices" in Chapter 1. This should get you thinking in terms of what needs to take place for your monitoring initiative to be successful, such as how to go about implementing, who to involve, and what pitfalls to avoid.

Chapter 2, "Theory of Operations," builds on Chapter 1's general design guidance by providing a theoretical overview of Nagios from the ground up. Rather than inundating you with configuration minutiae, Chapter 2 will give you a detailed understanding of how Nagios works without being overly specific about configuration directives. This knowledge will go a long way toward making configuration more transparent later.

Before we can configure Nagios to monitor our environment, we need to install it. Chapter 3, "Installing Nagios," should help you install Nagios, either from source or via a package manager.

Chapter 4, "Configuring Nagios," is the dreaded configuration chapter. Configuring Nagios for the first time is not something most people consider to be fun, but I hope I've kept it as painless as possible by taking a bottom-up approach, documenting only the most used and required directives, providing up front examples, and specifying exactly what objects refer to what other objects and how.

Most people who try Nagios become attached to it[2] and are loathe to use anything else. But if there is a universal complaint, it is certainly configuration. Chapter 5, "Bootstrapping the Nagios Config Files," takes a bit of a digression to document some of the tools available to make configuration easier to stomach. These include automated discovery tools, as well as graphical user interfaces.

In Chapter 6, "Watching: Monitoring Through the Nagios Plug-ins," we are finally ready to get into the nitty-gritty of watching systems, including specific examples with Nagios plug-in configuration syntax solving real-world problems. I begin with a section on watching Microsoft Windows boxes, followed by a section on UNIX, and ending with the "other stuff" section, which encompasses networking gear and environmental sensors.

Chapter 7, "Scaling Nagios," is new to the second edition. Scaling Nagios for large networks has been one of the most interesting problems Nagios sysadmins have had to deal with over the past five or six years. The explosion of machine virtualization and cost-effective cloud services have created a lot of interest in large parallel processing architectures that are composed of lots of little nodes. In this chapter, I cover several tools and strategies that will enable you to distribute the monitoring load and build a stable large-scale monitoring infrastructure for tens of thousands of nodes and beyond.

Chapter 8, "Visualization," covers one of my favorite topics: data visualization. Good data visualization solves problems that couldn't be solved otherwise, and I'm excited about the options that exist now, as well as what's on the horizon. With fantastic visualization tools like RRDTool, Ganglia, and Graphite, graphing time series data from Nagios is getting easier every day, but this chapter doesn't stop at mere line graphs.

Also new in the second edition is Chapter 9, "Nagios XI," which is dedicated to the new commercial version of Nagios. Built from many of the tools covered in this book by the guys who originally wrote Nagios, XI is truly a masterpiece of integration and usability. They've made monitoring with Nagios so simple my mom could do it (well, my mom writes optimizing cross-compilers for embedded FLIR systems, but you get my point).

And finally, now that you know the rules, it's time to teach you how to break them. At the time of this writing, Chapter 10, "The Nagios Event Broker Interface," is the only print documentation I'm aware of that covers the new Nagios Event Broker interface. The event broker is the most powerful Nagios interface available. Mastering it rewards you with nothing less than the ability to rewrite Chapter 2 for yourself by fundamentally changing any aspect of how Nagios operates or extending it to meet any need you might have. I describe how the event broker works and walk you through building an NEB module.

Who Should Read This Book?

If you are a systems administrator with a closet full of UNIX and Windows systems and assorted network gadgetry, and you need a world-class monitoring system on the cheap, this book is for you. Contrary to what you might expect, building monitoring systems is not a trivial undertaking. Constructing the system that potentially interacts with every TCP-based device in your environment requires a bit of knowledge on your part. But don't let that give you pause; systems monitoring has taught me more than anything else I've done in my career and, in my experience, no matter what your level of knowledge, working with monitoring systems has a tendency to constantly challenge your assumptions, deepen your understanding, and keep you right on the edge of what you know.

To get the most out of this book, you should have a pretty good handle on the text-based Internet protocols that you use regularly, such as SMTP and HTTP. Although it interacts with Windows servers very well, the Nagios daemon is meant to run on Linux, which makes the text pretty Linux heavy, so a passing familiarity with Linux or POSIX-ish systems is helpful. Although not strictly required, you should also have some programming skills. The book has a fair number of code listings, but I've tried to keep them as straightforward and as easy-to-follow as possible. With the exception of Chapter 8, which is exclusively C, the code listings are written in either UNIX shell or Perl.

Perhaps the only strict requirement is that you approach the subject matter with a healthy dose of open curiosity. If something seems unclear, don't be discouraged; check out the online documentation, ask on the lists, or even shoot me an email; I'd be glad to help if I can.

Have fun!

—Dave

End Notes

[1] www.portknocking.org

[2] Dare I say, love it?

## CHAPTER 1

Best Practices

Building a monitoring infrastructure is a complex undertaking. The monitoring system can potentially interact with every other system in the environment, and its consumers range from the layman to the highly technical. Building the monitoring infrastructure well requires not just technical aptitude, but also careful planning, a global perspective, and good people skills.

Perhaps most important, building monitoring systems also requires a light touch. An important distinction between good monitoring systems and bad ones is the amount of impact they have on network and server utilization, security, and the people entrusted with keeping things running. "Primum non nocere" isn't just for doctors; if you're building a monitoring infrastructure, it applies to you, too!

The first chapter in this book contains a collection of advice, gleaned from mailing lists such as the nagios-users list, other sysadmins, and hard-won experience. My hope is that this chapter will help you make some important design decisions up front, avoid some common pitfalls, and ensure that the monitoring system you build becomes a huge asset instead of a huge burden.

A Procedural Approach to Systems Monitoring

Good monitoring systems are not stacked up like a house of cards one script at a time by admin in separate silos. They are methodically created by admin with the support of their management and a clear understanding of the environment—both procedural and computational—within which they operate.

Without a clear understanding, for example, of which systems are considered critical, the monitoring initiative is doomed to failure. It's a simple question of context and usually plays out something like this:

Manager: I need to be added to all the monitoring system alerts.

Admin: All of them?

Manager: Well, yes; all of them.

Admin: Er, okay.

-next day-

Manager: My pager kept me up all night. What does this all mean?

Admin: Well /var filled up on server1 and the VPN tunnel to site5 was up and down.

Manager: Can't you just notify me of the stuff that's an actual problem?

Admin: Those *are* actual problems.

Certifications such as HIPPA, Sarbanes-Oxley, PCIDSS, and SSAE16 are requiring institutions such as universities, hospitals, and corporations to master the procedural aspects of their IT. This has had good consequences, because most organizations of any size today have contingency plans in place in the event that something bad happens. Disaster recovery, business continuity, and crisis planning attempt to ensure that the people in the trenches know what systems are critical to their business, and that they understand the steps to take to protect those systems in times of crisis, or recover them should they be destroyed. These certifications also seek to ensure that management has done due diligence to prevent failures to critical systems, for example, by installing redundant systems or moving tape backups offsite.

For whatever reason, monitoring systems seem to have been left out of this procedural approach to contingency planning. Most monitoring systems come into the network as a pet project of one or two small tech teams who have a specific need for it. Often, many teams employ their own monitoring tools independent of, and/or oblivious to, other monitoring initiatives going on within the organization. There seems no need to involve anyone else. Although this single-purpose approach to systems monitoring may solve an individual's or small group's immediate need, the organization as a whole suffers, and fragile monitoring systems always grow from it.

To understand why, consider that in the absence of a procedurally implemented monitoring framework, hundreds of critically important questions are nearly impossible to answer. Some examples follow:

- What amount of overall bandwidth is used for systems monitoring?
- What client-side UIDs are required to keep the monitoring system(s) running?
- What routers or other systems are they dependent on?
- Is sensitive information being transmitted in cleartext between hosts and the monitoring system?

If it was important enough to write a script to monitor a process, then it's important enough to consider what happens when the system running the script goes down, or when the person who wrote the script leaves and that userid is disabled. The piecemeal approach is by far the most common way monitoring systems are created, and yet the problems that arise from it are too many to be counted.

The core issue in our previous example is that there are no criteria that coherently define what a "problem" is, because these criteria don't exist when the monitoring system has been installed in a vacuum. Our manager felt that he had no visibility into system problems, and when provided with detailed information, still gained nothing of significance. This is why a procedural approach is so important. Before they do anything at all, the people undertaking the monitoring project should understand which systems in the organization are critical to the organization's operational well-being, and what management's expectation is regarding the uptime of those systems.

Given these two things, policy can be formulated that details support and escalation plans. Critical systems should be given priority, and their requisite pieces defined. That's not to say that the admin in our example should not be notified when /var is full on server1; just that when he is notified of it, he has a clear idea of what it means in an organizational context. Does management expect him to fix it now or in the morning? Who else was notified in parallel? What happens if he doesn't respond? This helps our manager as well. By clearly defining what constitutes a problem, management has some perspective on what types of alerts to ask for, and possibly more important, when they can go back to sleep.

Smaller organizations where there may only be a single part-time sysadmin are especially susceptible to piecemeal monitoring pitfalls. Thinking about operational policy in a four-person organization may seem like a silly waste of time, but in small environments, incident response is all the more important. A well-built monitoring system that enforces a well-thought out policy will enable a rapid and systematic response to problems—an aptitude no small shop can survive without.

Ideally, a monitoring system should enforce organizational policy rather than merely reflecting it. If management expects all problems on server1 to be looked at within 10 minutes, the monitoring system should provide our admin with a clear indicator in the message (such as a priority level), a mechanism to acknowledge the alert, and an automatic escalation to someone else at the end of the 10-minute window.

So how do we go about finding out what the critical systems are? Senior management is ultimately responsible for the overall well-being of the organization, so they should be the ones making the call. This is why management buy-in is so vitally important. If you think this is beginning to sound like disaster-recovery planning, you're ahead of the curve. Disaster recovery works toward identifying critical systems for the purpose of prioritizing their recovery; therefore, it is a methodologically identical process to planning a monitoring infrastructure. In fact, if a disaster recovery plan already exists, that's the place to begin. The critical systems have already been identified.

Critical systems as outlined by senior management will not be along the lines of "all problems with server1 should be looked at within 10 minutes." They'll probably be defined as logical entities: for example, "Email is critical." So after the critical systems have been identified, the implementers will dissect them one by one into their parts. Be sure to involve all interested parties in this process. Email administrators will have a good idea of what email is composed of, and criteria for the monitoring of email, which, if not met, will inevitably result in them rolling their own monitoring tools.

Work with as many teams as possible to get, if not a solution that works for everyone, at least specific requirements for what it means to monitor every critical system. Where custom monitoring scripts already exist, don't dismiss them; instead, try to incorporate them. Groups tend to trust the tools they're already using, so co-opting those tools usually buys you some support. Nagios does an excellent job at using external monitoring logic along with its own scheduling and escalation rules.

Processing and Overhead

Monitoring systems necessarily introduce some overhead in the form of network traffic and resource utilization on the monitored hosts. Most monitoring systems typically have a few specific modes of operation, so the capabilities of the system, along with implementation choices, dictate how much overhead is introduced and where.

Remote Versus Local Processing

Nagios exports service checking logic into tiny single-purpose programs called plug-ins. This makes it possible to add checks for new types of services quickly and easily, as well as co-opt existing monitoring scripts. This modular approach makes it possible to execute the plug-ins themselves, either locally on the monitoring server or remotely on the monitored hosts.

Centralized execution is generally preferable whenever possible because the monitored hosts bear less of a resource burden, and all the configuration resides in a single place. However, remote processing may be unavoidable or even preferred in some situations. For very large environments with tens of thousands of hosts, centralized execution may be too much for a single monitoring server to handle. In this case, the monitoring system may need to rely on the clients to run their own service checks and report back the results. Some types of checks may be impossible to run from the central server at all. For example, plug-ins that check the system load or amount of free memory may require remote execution.

At the risk of confusing things, I should mention that there are quite a few ways to blur the line between local and remote. This includes plug-ins like Check_MK (covered in Chapter 6, "Watching: Monitoring through the Nagios Plug-ins"), which implement "super checks"—single-service checks that return results for multiple services. Included also in this list are various distributed monitoring schemes (covered in Chapter 7, "Scaling Nagios"), which spread the service check load over multiple Nagios servers, and integration with other systems like Ganglia (covered in Chapter 8, "Visualization"), which enables state-sharing and the reuse of check results from sister monitoring systems.

Bandwidth Considerations

Nearly all plug-ins will generate some IP traffic. Each network device this traffic must traverse introduces network overhead, as well as a dependency into the system. In Figure 1.1, there is a router between the Nagios Server and Server1. Because Nagios must traverse the router to connect to Server1, Server1 is said to be a child of the router. It is always desirable to do as little layer 3 routing between the monitoring system and its target hosts as possible, especially where devices like firewalls and WAN links are concerned. So the location of the monitoring system (or systems) within the network topology becomes an important implementation detail.

In addition to minimizing layer 3 routing of traffic from the monitoring host, we also want to make sure that the monitoring host is sending as little traffic as possible. This means paying attention to things like polling intervals and plug-in redundancy. Plug-in redundancy occurs when two or more plug-ins are effectively monitoring the same service.

Redundant plug-ins may not be obvious. They usually take the form of two plug-ins that measure the same service, but at different depths. Take, for example, an imaginary web service running on server1. The monitoring system may initially be set up to connect to port 80 of the web service to see if it is available. Then some months later, when the website running on server1 has some problems with users being able to authenticate, a plug-in may be created that verifies authentication is working correctly. All that is actually needed in this example is the second plug-in. If it can log in to the website, then port 80 is obviously available, and the first plug-in is doing nothing but wasting resources. Plug-in redundancy may not be a problem for smaller sites. But small sites have a habit of turning into large sites, and for large sites, eliminating plug-in redundancy (or better, ensuring it never occurs in the first place) can greatly reduce the burden on the monitoring system and the network.

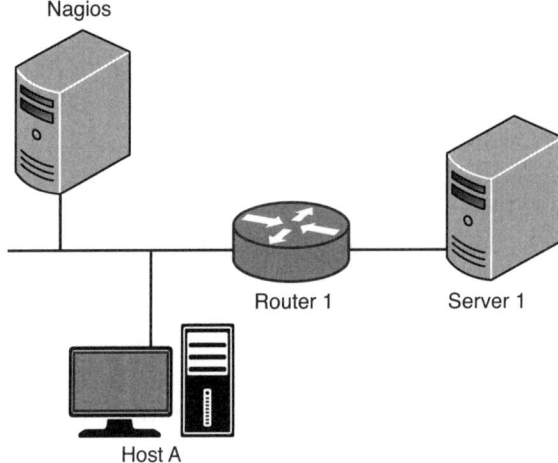

Figure 1.1 The router between Nagios and server1 introduces a dependency, and some network overhead in the form of layer 3 routing decisions.

Minimizing the overhead incurred on the environment as a whole means maintaining a global perspective on its resources. Hosts connected by slow WAN links that are heavily utilized, or are otherwise sensitive to resource utilization, should be grouped logically. Nagios provides "hostgroups" for this purpose. These allow configuration settings to be optimized to meet the needs of the group. For example, plug-ins may be set to a higher timeout for the "Remote-Office" hostgroup, ensuring that network latency doesn't cause a false alarm for hosts on slower networks. Special consideration should be given to the location of the monitoring system to reduce its impact on the network, as well as to minimize its dependency on other devices (discussed later in this chapter). Finally, make sure that your configuration changes aren't needlessly increasing the burden on the systems and network you are monitoring, as in the case of redundant plug-ins. The last thing a monitoring system should do is cause problems of its own.

Network Location and Dependencies

The location of the monitoring system within the network topology has wide-ranging architectural ramifications, so you should take some time to mull it over. Your implementation goals are threefold:

1. Maintain existing security measures.

2. Minimize impact on the network.

3. Minimize the number of dependencies between the monitoring system and the most critical systems.

No single ideal solution exists, so these three goals need to be weighed against each other for every environment. The end result will always be a compromise, so it's important to spend some time diagramming out a few different architectures and considering the consequences of each.

The network topology shown in Figure 1.2 is a simple example of a network that should be familiar to any sysadmin. Most private networks today that provide Internet-facing services have at least three segments: the inside, the outside, and the DMZ. In our network example, the greatest number of hosts exist on the inside segment. Most of the critically important[1] hosts, however, exist on the DMZ.

Acme Web Hosting Company

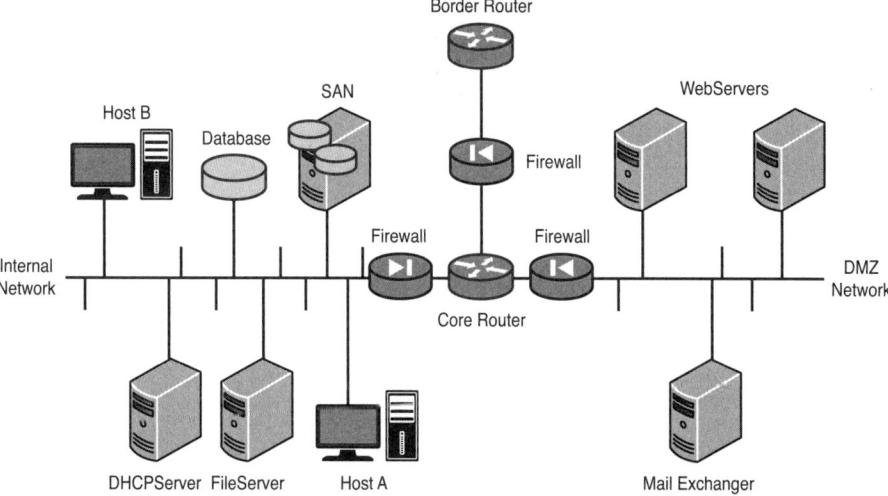

Figure 1.2   A typical two-tiered network

Following the implementation rules at the beginning of this section, our first priority is to maintain the security of the network. Creating a monitoring framework will necessitate that some ports on the firewalls be opened so that, for example, our monitoring host can connect to port 80 on hosts in other network segments. If the monitoring system were placed in the DMZ, many more ports on the firewalls would need to be opened than if the monitoring system were placed on the inside segment, simply because there are more hosts on the internal segment. For most organizations, placing the monitoring server in the DMZ would be unacceptable for this reason. Security is discussed in more depth in the next section, but for this example, it's simple arithmetic.

There are many ways to reduce the impact of the monitoring system on the network. For example, the use of a "FoxBox" or similar dedicated GSM or CDMA device to send SMS messages reduces network traffic and removes dependencies. In the context of network location, however, the best way to minimize network impact is to place the monitoring system on the segment with the largest number of hosts, because this ensures that less traffic must traverse the firewalls and router. This again points to the internal network.

Finally, placing our monitoring system in a separate network segment from most of the critical systems is not ideal, because in the event that one of the network devices becomes unavailable, the monitoring system loses visibility to the hosts behind it. Nagios refers to this as a network blocking outage. The hosts on the DMZ are "children" of their firewall, and when configured as such, Nagios is aware of the dependency. In the event that the firewall goes down, Nagios does not have to send notifications for all the hosts behind it (but it can if you want it to), and the status of those hosts will be flagged as "unknown" in availability reports for the amount of time that they were not visible. Every network will have some amount of dependency, so this needs to be considered in the context of the other two goals. In our example, despite the dependency, the inside segment is probably the best place for the monitoring host.

Security

The ease with which large monitoring systems can become large rootkits makes it imperative that security is considered sooner rather than later.

Because a monitoring system usually needs remote execution rights to the hosts it monitors, it's easy to introduce backdoors and vulnerabilities into otherwise secure systems. Worse, because they're installed as part of a legitimate system, these vulnerabilities may be overlooked by penetration testers and auditing tools. The first and most important thing to look for when building secure monitoring systems is how remote execution is accomplished.

Historically, commercial monitoring tools have included huge monolithic agents, which must be installed on every client to enable even basic functionality. These agents usually include remote shell functionality and/or proprietary bytecode interpreters, which allow the monitoring host carte blanche to execute anything on the client via its agent. This implementation makes it difficult at best to adhere to basic security principles, such as least privilege. Anyone with control over the monitoring system has complete control over every box it monitors.

Nagios, by comparison, follows the UNIX adage: "Do one thing and do it well." It is really nothing but a task optimized scheduler and notification framework. It doesn't have an intrinsic capability to run programs on other computers, and it contains no agent software at all. These functions exist as separate, single-purpose programs that Nagios must be configured to use. By outsourcing remote execution to external programs, Nagios maintains an off-by-default policy and doesn't attempt to reinvent things like encryption protocols, which are critically important, and difficult to implement. With Nagios, it's simple to limit the monitoring servers access to its clients, but poor security practices on the part of admin can still create insecure systems. In the end, security is always up to you.

The monitoring system should have only the access it needs to remotely execute the specific plug-ins required. Avoid rexec-style plug-ins that take arbitrary strings and execute them on the remote host. Ideally, every remotely executed plug-in should be a single-purpose program, which the monitoring system has specific access to execute. Some very useful plug-ins provide lots of functionality in a single binary. Check-MK I've already mentioned. NSCLIENT++ for Windows can query any perfmon counter on a remote system. Multipurpose super plug-ins are too compelling not to use in many environments, but you should, at a minimum, make use of their authorization features to ensure that access is as limited as possible.

The communication channel between the remotely executed plug-in and the monitoring system should be encrypted. Avoid nonstandard or proprietary encryption protocols.[2] Encryption protocols are notoriously difficult to implement, let alone create. The popular remote execution plug-ins for Nagios use the industry-standard OpenSSL library, which is peer reviewed constantly by smarter people than you and I. Even if none of the information being passed is considered sensitive, the implementation should include encrypted channels from the get-go as an enabling step. If the system is implemented well, it will grow fast, and it's far more difficult to add encrypted channels after the fact than it is to include them in the initial build.

Simple Network Management Protocol (SNMP), a mainstay of systems monitoring that is supported on nearly every computing device in existence today, should not be used on public networks, and should be avoided if possible on private ones. For most purposes involving general-purpose workstations and servers, far better alternatives to SNMP exist. If SNMP must be used for network equipment, try to use SNMPv3, which includes encryption. No matter what version you use, make sure it's configured in a read-only capacity and accepts connections only from specific hosts. For whatever reason, sysadmins seem chronically incapable of changing SNMP community string names. This simple implementation flaw accounts for most of SNMP's bad rap. Look for more info on SNMP in Chapter 6, "Watching: Monitoring Through the Nagios Plug-ins."

Some organizations have network segments that are physically separated or otherwise inaccessible from the rest of the network. In this case, monitoring services on the isolated subnet means adding a Network Interface Card (NIC) to the monitoring server and connecting it to the private segment, or making the monitoring server VLAN aware and placing it on a trunk switchport. Isolated network segments are usually isolated for a reason, so at a minimum, the monitoring system should be configured with strict local firewall rules preventing them from forwarding traffic from one subnet to the other. Consideration should be paid to building separate monitoring systems for nonaccessible networks.

When holes must be opened in the firewall for the monitoring server to check the status of hosts on a different segment, consider using remote execution to minimize the number of ports required. For example, the Nagios Box in Figure 1.3 must monitor the web server and SMTP daemon on Server1. Instead of opening three ports on the firewall, the same outcome may be reached by running a service checker plug-in remotely on Server1 to check that the apache and qmail daemons are running. By opening only one port instead of three, less opportunity exists for abuse by a malicious party. Super plug-ins like check_mk do an excellent job of tunneling a lot of service checks over a single connection to a single port.

Scenario 1: Nagios runs local plugins to check port availability. Three firewall rules required.

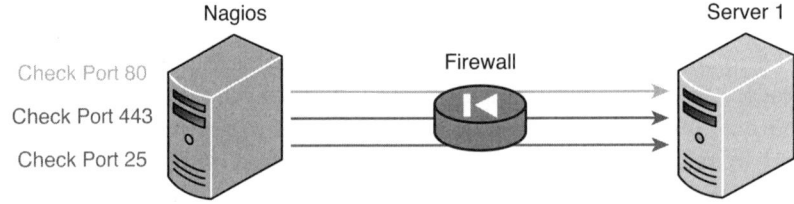

Scenario 2: Nagios uses remote execution to check if the services are running. One firewall rule required.

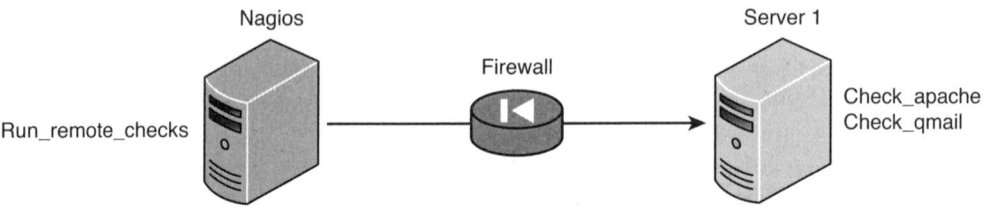

Figure 1.3   When used correctly, remote execution can enhance security by minimizing firewall ACLs.

A good monitoring system does its job without creating flaws for intruders to exploit, and Nagios makes it simple to build secure monitoring systems as long as the implementers are committed to building them that way.

Silence Is Golden

With any monitoring system, a balance must be struck between too much granularity and too little. Technical folk, like sysadmins, usually err on the side of offering too much. Given 20 services on 5 boxes, many sysadmins monitor everything and get notified on everything, whether the notifications might represent a problem or not.

For sysadmins this is not a big deal; they have a tendency to develop an organic understanding of their environments, and the notifications serve as an additional point of visibility, or as an event correlation aid. For example, a notification from workstation1 that its network traffic is high, combined with a CPU spike on router 12 and abnormal disk usage on server3, may indicate to a sysadmin that Ted from accounting has come back early from vacation. A diligent sysadmin might follow up on that hunch to verify it really is Ted and not a teenager at the University of Hackgrandistan owning Ted's workstation. For the non-sysadmin, however, the most accurate phrase to describe these notifications is: "false alarm."

Typically, monitoring systems use static thresholds to determine the state of a service. The CPU on server1, for example, may have a threshold of 95%. When the CPU goes above that, the monitoring system will send notifications or perform some automatic break/fix. One of the biggest mistakes an implementer can make when introducing a monitoring system into an environment is simply not having taken the time to find out what the normal operating parameters on the servers are. If server1 typically has 98% CPU utilization from 12 a.m.– 2 a.m. because it does batch processing during these hours, a false alarm will be sent.

False alarms should be methodically hunted down and eradicated. Nothing can undermine the credibility of, and erode support for, a fledgling monitoring system like people getting what they think are useless false alarms. Before the monitoring system is configured to send notifications, it should be run for a few weeks to collect data on at least the critical hosts to determine what their normal operational parameters are. This data, collectively referred to as a *baseline*, is the only reasonably responsible way to determine static thresholds for your servers.

That's not to say our sysadmin should be prevented from getting the most out of his cell phone's unlimited data plan. I'm merely suggesting that some filtering be put in place to ensure no one else need share his unfortunate fascination. One great thing about following the procedural approach outlined earlier in this chapter is that it makes it possible to think about the organization's requirements for a particular service on a specific host *before* the thresholds and contacts are configured. If Alice the DBA doesn't need to react to high CPU on server1, then she should not get paged about it.

Nagios provides plenty of functionality to enable sysadmins to be notified of "interesting events" without alerting management or other noninterested parties. With two threshold levels (warning and critical) and a myriad of escalation, group, and polling options, it is relatively simple to get "early and often" style notifications for control freaks, while keeping others abreast of "just the problems." It is highly recommended that a layered approach to notification be a design goal of the system from the beginning.

Good monitoring systems tend to be focused rather than chatty. They may monitor many services for the purpose of historical trending, but they send fewer notifications than one would expect, and when they do, it's to the group of people who want to know. For the intellectually curious who don't want their pagers going off at all hours of the day and night, consider sending summary, or digest, reports every 24 hours or so. Nagios has some excellent built-in reporting functionality, which makes it easy to provide historical availability and notification reports on predefined groups of systems.

Watching Ports Versus Watching Applications

In the "Processing and Overhead" section earlier in the chapter, we briefly discussed redundant plug-ins that monitored a web server. One plug-in connected to port 80 on the web server, while the other attempted to log in to the website hosted by the server. The latter plug-in is an example of what is sometimes referred to as *End to End (E2E)* or application-layer monitoring. E2E monitoring makes use of the monitored services in the same way a user might. Instead of monitoring port 25 on a mail server, the E2E approach would be to send an email through the system. Instead of monitoring the processes required for CIFS, an E2E plug-in would attempt to mount a shared drive, and so on.

Although introducing more overhead individually, E2E plug-ins can lighten the load when used to replace several of their conventional counterparts. A set of plug-ins that monitors a web application by checking the web ports, database services, and application server availability might be replaced by a single plug-in that logs in to the website and makes a query. E2E plug-ins tend to be "smarter." That is, they catch more problems by interpreting the outcome of an attempted use of a service rather than watching likely points of failure. For example, an E2E plug-in that parses the content of a website will find an alert on a permissions problem, whereas a simple port watcher will not.

Sometimes that's a good thing, and sometimes it isn't. What E2E gains in rate of detection, it loses in resolution. What I mean by that is, with E2E, you often know that there is a problem, but not where the problem actually resides, which can be bad when the problem is in a completely unrelated system. For example, an E2E plug-in that watches an email system will detect failure and send notifications in the event of a DNS outage because the mail servers cannot perform MX lookups and therefore cannot send mail. This makes E2E plug-ins susceptible to what some may consider false alarms, so they should be used carefully.

On the other hand, a problem in some unrelated infrastructure, which is affecting a system responsible for transferring funds, is something the bank management will want to know about regardless of root-cause. E2E is great at catching failures in unexpected places and can be a real lifesaver when used on systems where problem detection is absolutely critical. This is yet another reason to identify those critical systems at the onset.

Adoption of E2E is slow among the commercial monitoring systems because it's difficult to predict what customers' needs are, and that makes writing turnkey agent software difficult at best. Nagios, by comparison, excels at this sort of application-layer monitoring. It makes no assumptions about how you want to monitor your environment, which simplifies things when you want to extend its functionality. I'll talk a lot more about plug-ins and how they work in Chapter 2, "The Nagios Systems Monitoring Theory of Operations."

Who's Watching the Watchers?

If there is a fatal flaw in the concept of systems monitoring, it is the use of untrustworthy systems to watch other untrustworthy systems. In the event that your monitoring system fails, it's important you are at least informed of it. A failover system to pick up where the failed system left off is even better.

The specifics of your network dictate what needs to happen when the monitoring system fails. If you are bound by strict SLAs, then uptime reports are themselves a critical part of your business, and a failover system should be implemented. Often it's enough to know that the monitoring system is down.

Failure proofing monitoring systems is a messy business. Unless you work at a tier1 ISP, you'll always hit some upstream dependency you have no control over if you go high enough into the topology of your network. This does not negate the necessity of a plan.

Small shops should at least have a secondary system, like a syslog box, or some other piece of infrastructure that can heartbeat the monitoring system and send an alert in the event that things go wrong. Large shops may want to consider global monitoring infrastructure, either provided by a company that sells such solutions or by maintaining a mesh topology of hosted Nagios boxes in geographically dispersed locations.

Nagios makes it easy to mirror state and configuration information across separate boxes. Configuration and state are stored as terse, cleartext files by default. Configuration syntax hooks make event mirroring a snap, and Nagios can be configured in distributed monitoring scenarios with multiple Nagios servers. The monitoring system may be the system most in need of monitoring; don't forget to include it in the list of critical systems.

End Notes

[1] Critically important as inferred from the fact that this is a web hosting company, and those are web servers.

[2] A common mistake among commercial monitoring apps.

## CHAPTER 2

Theory of Operations

Monitoring servers, as a consequence of their job, interact with other systems quite a bit. They speak SMTP to the mail systems, SQL to the database systems, and HTTP, HTTPS, AJP, SOAP, and by the time you read this, who knows what else to the web servers. The monitoring server could be considered an odd sort of multilayer protocol simulator. It needs to be able to give input to, and understand the output of, every protocol spoken by every system in any environment.

Imagine for a moment, being tasked with creating such a system. How would you build it? Would you brute-force it? Even though there are a lot of protocols, they are finite, so it should be possible to create one big program that can monitor every conceivable gadget that anyone could ever want to monitor, right? This seems to be the thinking behind the large commercially available monitoring systems I've used. Some stop there, and others, perhaps sensing in some small way the monumental hubris of the assumption that no one might ever require something that they failed to consider, consider what might happen if someone did. When this happens, they provide for you—as if it were a gift—a proprietary, domain-specific scripting language that runs in a proprietary interpreter, which is embedded in their monolithic monitoring software.

As you can imagine, this approach presents a few problems. The complexity of the software may be the single largest impact. Many large monitoring packages have GUIs with menus 10 to 15 selections deep. The agent software becomes bloated fairly quickly and is often larger than a 500Mb per server. Security is difficult to manage because the monitoring program assumes you want the entire feature set available on every monitored host, and this makes it difficult to limit the monitoring server's access to its clients. The package, as a whole, is only as good as the predictions of the developers and marketeers behind it. Finally, the quite deliberate consequence of using proprietary scripting languages is vendor lock-in. It is a rare sysadmin who relishes the idea of porting several years of monitoring customizations to a general purpose language, or worse, into a different vendor's proprietary language. I've quit jobs over less.

Nagios, by comparison, takes the opposite approach. It has no internal monitoring logic, assumes next to nothing about what, or how, you might want to watch. It neither requires nor provides agent software, and it contains no built-in proprietary interpreters. In fact, Nagios isn't really a "monitoring application" at all, in the sense that it doesn't actually know how to monitor anything. So what *is* Nagios exactly, and how does it work?

This chapter should provide some insight into what Nagios does, how it goes about doing it, and why. Throughout the chapter, I mention various configuration options that are available in Nagios, in the context of subject matter, but this chapter is meant to provide a conceptual understanding of the mechanics of Nagios as a program. Chapter 4, "Configuring Nagios," will cover the configuration options in detail.

The Host and Service Paradigm

Nagios is an elegant program that is quite simple to understand. It does exactly what you would want, in a way that you would expect, and can be extended to do some amazing things. After you grasp a few fundamental concepts, you will feel completely empowered to go forth and build the monitoring system your Openview-burdened friends can only dream about.

Starting from Scratch

The easiest way to understand what Nagios is and what it does is to go back to our description of the piecemeal approach to systems monitoring in Chapter 1, "Best Practices." The piecemeal approach usually happens when a sysadmin has just been burned by an important service or application. The service in question has gone down, and the admin found out about it from his customers or manager, creating the perception that he's not aware of what's happening with his systems. Sysadmin are a proactive bunch, so before too long, our admin has a group of scripts that he runs against his servers. These scripts check the availability of various things. At least one of them looks something like this:

```
ping -qc 5 server1 || (echo "server1 is down" | mail dude@domain.org)
```

This shell script sends five ICMP echo packets to server1, and if server1 doesn't reply, it emails the sysadmin to notify him. This is a good thing. The script is easy to understand, can be run from a central location, and answers an important question. But, soon, bad things start to happen.

One day, the router between our admin's workstation and servers 1 through 40 goes down. Suddenly the network segment is no longer visible to the system running the scripts. This causes 40 emails to be needlessly sent to our admin, one for each server that is no longer pinging. Later, another administrator and a few managers want to get different subsets of these notifications, so our sysadmin creates a group of mailing lists. But some people soon get duplicate emails because they belong to more than one list, and each of those lists has received the same notification. Some weeks later, our admin gets a noncritical notification at 3 a.m. He decides to fix it in the morning and goes back to sleep. But when morning arrives, he forgets all about it. The service remains unavailable until a customer notices it and calls on the phone.

Our admin had the right notion, and wrote a script in a language that suited him, that effectively detected the outage. He doesn't need better scripts, just a smarter way to run them. He needs a task-efficient scheduling and notification system: a system that tracks the execution and status of a bunch of little monitoring scripts, manages dependencies between groups of monitored objects, provides escalation, and ensures people don't get duplicate pages, regardless of their memberships. That, in a nutshell, is Nagios.

Nagios is a framework to which you can entrust your own little monitoring programs written in any language you choose. Given a few configuration parameters, Nagios will handle the scheduling for you, and when each little monitoring program (or plug-in) reports its status back to Nagios, it will make sure the right people get notified and provide escalations, if necessary. It also keeps track of the dates and times that various plug-ins changed states and has nice built-in historical reporting capabilities. Nagios has lots of hooks that make it easy to get data in and out, so it can provide real-time data to graphing programs, such as RRDTool and MRTG, and can easily cooperate with other monitoring systems, either by feeding them or by being fed by them.

One of the things I like best about Nagios is that it leverages what you're already good at (individually and organizationally) and doesn't throw your hard work into the bit-bucket. If you are a TCL Jedi and your organization values you because of your skills, it shouldn't be forced to trash the five months you spent on a TCL-based monitoring infrastructure in an effort to better centralize their monitoring tools. Because Nagios has no desire to control your monitoring methodology, it won't attempt to drive your organization's use of tool sets and therefore will never force you to reinvent the wheel.

Hosts and Services

I mentioned earlier in the chapter that Nagios makes few assumptions about what and how you want to monitor. It allows you to define everything. Definitions are the bread and butter of how Nagios works. Every element that Nagios operates with is user defined. For example, Nagios knows that in the event a plug-in returns a critical state, it should send a notification, but Nagios doesn't know what it means to send one. You define the literal command syntax Nagios uses to notify contacts, and you may do this on a contact-by-contact basis, a service-by-service basis, or both. Most people use email notifications, and you'll find that there are existing definitions for most of the things you want Nagios to do, so you don't really *have* to define everything, but little of how Nagios works is written in stone.

The most important assumption Nagios makes about your environment is that there are hosts and services. These are the two most fundamental object types in Nagios. You may think of a host in terms of a physical entity. Servers and network appliances are the most common types of hosts, but really, a host is anything that speaks TCP. Certain types of environmental sensors and my friend Chris's refrigerator are also examples of hosts. Services are the logical entities that hosts provide. The web server daemon that runs on the server sitting in the rack is a service.

Typically, a single host runs multiple applications—or at least has multiple elements that bear watching—but the host will either be up or down, available or not. Therefore, Nagios allows you to define a single host check mechanism and multiple service checks for each host. These host and service check definitions are what tell Nagios which plug-ins to call to obtain the status of a host or service. For example, the check_ping plug-in may be defined as the host check for server1. If the host check fails, the host is not available. If the host is down, all the services on that host are also not available, so it would be silly to send a page for each individual service. It would be silly, in fact, to run the service checks at all, until the host itself becomes available again.

The hosts/services assumption makes it easy for Nagios to track which services are dependent on what hosts. When Nagios runs a plug-in on a service provided by a host and that plug-in returns an error state, the first thing Nagios will do is run the host check for that host. If the host check also returns an error state, the host is unavailable, and Nagios will notify you only of the host outage, postponing the service checks until the host becomes available again.

Interdependence

In fact, this idea of interdependence is pervasive throughout Nagios. Nagios tends to be smart about not wasting resources by running checks on, and sending notifications about, hosts and services that are obviously unavailable. Nagios tracks dependencies between services on different hosts two different ways.

The first is child/parent relationships, which may be defined only for hosts. Every host definition may optionally specify a parent using the parents directive. This works well for hosts behind network gear such as firewalls and even virtualized servers. If the parent of host1 goes down, Nagios will consider host1 unreachable, instead of down, which is an important distinction for people with a service level agreement to live up to. While a host is in an unreachable state, Nagios won't bother to run host or service checks for it. However, Nagios can be configured to send notifications for unreachable hosts if you want it to.

The second way Nagios can track dependencies between hosts is with dependency definitions. These definitions work for both hosts and services and are used to track more subtle dependency relationships between different hosts, or services running on different hosts. A good example of this type of dependency tracking is a web proxy. In Figure 2.1, the firewalls are configured so that only one host on the secure network is allowed to connect to the web server's port 80. All other servers must proxy their web requests through the proxy service on this server. Nagios is no exception; if it wants to check the status of the web server's port 80, it must do so through the proxy server. Because Nagios doesn't rely on the web proxy for any other type of network access, a parent/child relationship is not appropriate. What is needed is a way to make the web server's port 80 dependent on the web proxy's port 8080. If the web proxy service goes down, Nagios should not check on, nor notify, the web service. This is exactly what dependency relationships do.

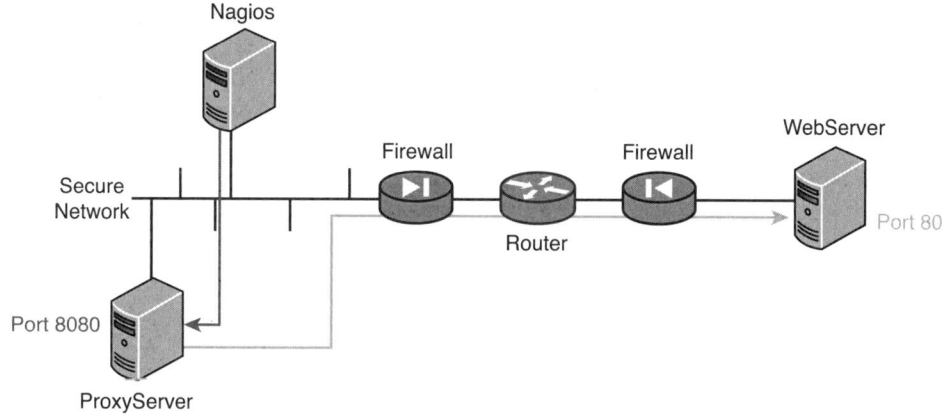

Figure 2.1   Nagios uses dependency definitions to track interdependent services.

The Downside of Hosts and Services

In my opinion, the manner in which Nagios naturally handles the host and services paradigm is genius. It is simple to understand, always does what you would expect, and makes things generally easy to manage. However, the hosts and services assumption also limits the functionality of Nagios to some degree.

To understand why, consider a large corporate or university email system. Such a system is composed of MXs and border mail systems, internal relay servers, and user-facing groupware. Outages of various services and hosts within the email system affect the entity as a whole, but don't necessarily make it completely unavailable. An MX outage, for example, could do nothing at all to affect the flow of mail,[1] whereas a groupware outage might mean that mail is still being delivered to the MXs, but that users cannot interact with it.

Business processes and higher-level entities, such as email, are difficult to capture on a host and service scale because they are an aggregation of many services on many hosts. Nagios provides host and service groups, which can contain individual services from different hosts. So a service group called email can be created, which would summarize the status of each service that corporate email depends upon. But given that post office protocol (POP) on server1 is unavailable, it is not obvious to the uninitiated what effect this outage has on the overall email entity.

Given that Nagios plug-ins are user defined, and in many cases, user created, an enterprising admin could write a single plug-in to check the overall status of the email system piece by piece. But within the hosts and services paradigm, to which host would the service that plug-in checks belong? The hosts and services assumption makes it a bit difficult to model larger entities, but that's not necessarily "game over" for those who need to do so. In later chapters, I'll cover several methods of incorporating business processes using your existing checks. These include the check_mk plug-in, the business process views built in to Nagios XI, and some visualization options.

Plug-ins

We've discussed that Nagios is a scheduling and notification framework that calls small monitoring programs called plug-ins. Next we look at what these plug-ins are and how they interact with Nagios to provide a fully featured monitoring system.

Exit Codes

The first thing you should probably know is that you don't have to write your own monitoring system from scratch. The Nagios-Plugins project[2] is a collection of user-contributed plug-ins that contains the functionality you would expect from any monitoring system. These include plug-ins that run on the monitoring system, such as port scanners and

ICMP and SNMP query tools, as well as those designed to be executed remotely, such as CPU and memory utilization checkers. In addition to those included in the plug-ins project, thousands of special-purpose plug-ins are available from the Nagios Exchange at www.nagiosexchange.com.

Eventually you will want either to write your own plug-in or to reuse some existing scripts as Nagios plug-ins, and doing so couldn't be easier. A Nagios plug-in is nothing but a program that provides a specific exit code, as defined in Table 2.1.

Table 2.1   *Nagios Plug-in Exit Codes*

Code	Meaning
0	Okay
1	Warning
2	Critical
3	Unknown

Providing an exit code is, literally, the sole requirement by which a plug-in must abide. Any programming or scripting language that can provide an exit code[3] can be used to write a Nagios plug-in. A plug-in's job is usually to do the following:

1. Grab some bit of information from a host, such as its current load or its index.html page.
2. Compare that bit of information against an expected state or threshold.
3. Provide an exit code describing the outcome of that comparison.

Plug-ins are specified in service definitions, along with other details, such as the name of the service and how often to execute a check. Nagios handles the scheduling and execution of the plug-in per the service definition and optionally provides it with thresholds to compare against. After the plug-in does its "thing," it returns one of four exit codes: 0 for "okay," 1 for "warning," 2 for "critical," or 3 for "unknown." Nagios parses the plug-in's exit code and responds accordingly. In the event Nagios receives a bad code, it updates the service state and may or may not contact people, but I cover more on that later.

Listing 2.1 shows our ping shell script from the "starting from scratch section," rewritten as a Nagios plug-in.

Listing 2.1 *A Ping Plug-in*

```
#!/bin/sh
if ping -qc 5 server1
then
  exit 0
else
    exit 2
fi
```

Our ping command still sends five ICMP packets to server1, but this time it exits 0 if the command is successful and 2 if it is not.

Traditionally, in addition to the exit code, plug-ins also provide a single line of text on stdout which Nagios will interpret as a human-readable summary. This text is made available, verbatim, to the web interface, which will display it in a status field where appropriate. This is handy for passing back information about the service to humans, who aren't big on parsing exit codes. Listing 2.2 is our ping plug-in, now modified to give Nagios some summary text.

Listing 2.2 *Ping with Summary Output*

```
#!/bin/sh

OUTPUT='ping -c5 server1 | tail -n2'
If [ $? -gt 0 ]
then
    echo "CRITICAL!! $OUTPUT"
    exit 2
else
    echo "OK! $OUTPUT"
    exit0
fi
```

This time we use the $OUTPUT variable, combined with the tail command, to capture the last two lines of the ping command. Now, when the service is viewed in Nagios' spiffy web interface, something such as the following text will appear in the Status Information field:

```
5 packets transmitted, 5 packets received, 0% packet loss round-trip
min/avg/max = 0.1/0.8/3.9 ms
```

In Nagios 3.0 and later, plug-ins may pass back more than a single line of text. Nagios will still provide the first line to the web interface and will store the subsequent lines in an internal data structure called a "macro."

One of the many nice things about the way the Nagios plug-in architecture works is that, because every plug-in is a little self-contained program, it is possible to launch them from the command line. The plug-in development guidelines specify that every plug-in should have an -h, or help switch, so if you need to find out how a plug-in works, you can usually call it from the command line with -h. This also makes troubleshooting plug-in problems a snap. If Nagios is having trouble with a plug-in, you can execute the plug-in directly from a shell prompt with the same arguments and see what's happening.

That is all you need to know to modify your existing scripts for use with Nagios. Of course, if you really want to get serious about writing plug-ins that other people might want to use, you should check out the plug-in development guidelines available from the plug-in development team at:

```
http://nagiosplug.sourceforge.net/developer-guidelines.html
```

Remote Execution

In the "Processing and Overhead" section in Chapter 1, and several other times in the preceding text, I've made reference to plug-ins that either run locally on the Nagios server or remotely on the monitored hosts. Given that Nagios has no means of carrying out remote execution, it's important to understand some of the various methods by which it is accomplished in practice.

The easiest way to understand how Nagios launches plug-ins on remote servers is to revisit our sysadmin, now familiar from the previous examples. When he has a need to perform remote execution, he turns to SSH. Let's say, for example, that he wants to query the load average of a remote system. This is accomplished easily enough:

```
$ ssh server1 "uptime | cut -d: -f4"
```

The SSH client launches the command uptime | cut -d: f4 on the remote server and passes back the output to the local client (in our example, this would be something like 0.08, 0.02, 0.01). This is fine, but our sysadmin wants something that will page him if the 15-minute average is above three, so he writes the script in Listing 2.3 and places it on the remote server.

Listing 2.3 *A Remote Load Average Checker*

```
#!/bin/sh

LOAD='uptime | awk '{print $12}'
```

```
if [ $LOAD -gt 1 ]
then
    echo "high load on 'hostname' | mail dude@domain.org"
fi
```

In this script, the output of uptime is filtered through awk, which extracts the last number from uptime's output. This number happens to be the 15-minute load average. This number is compared against 1, and if it is greater, our admin receives an email. After this script is saved as load_checker.sh and placed in /usr/local/bin on the remote server, our admin can execute it with SSH remotely, like so:

```
ssh server1 "/usr/local/bin/load_checker.sh"
```

In reality, he'd probably just schedule it in cron on the remote box, but bear with me for a moment. An interesting thing about executing scripts remotely with SSH is that not only does SSH capture and pass back the output from the remote script, but also its *exit code*.

The upshot is that if our sysadmin were to rewrite his script to look like the one in Listing 2.4, he would have a Nagios plug-in.

Listing 2.4 *A Remote Load Average Checker with Exit Codes*

```
#!/bin/sh

LOAD='uptime | awk '{print $12}'
if [ $LOAD -gt 1 ]
then
    echo "Critical! load on 'hostname' is $LOAD"
    exit 2
else
    echo "OK! Load on 'hostname' is $LOAD"
    exit 0
fi
```

But how does Nagios execute the remote command, and capture its output and code, when it can only execute programs on the local hard drive? Simple: We just write a local plug-in around our sysadmin's SSH command. This script might look something like Listing 2.5.

Listing 2.5   *A Script That Calls load_checker and Parrots Its Output and Exit Code*

```
#!/bin/sh

#get the output from the remote load_checker script
OUTPUT='ssh server1 "/usr/local/bin/load_checker.sh"'

#get the exit code
CODE=$?

echo $OUTPUT
exit $CODE

fi
```

The script in Listing 2.5 doesn't have any conditional logic. Its only job is to execute the remote script and parrot back its output and exit code to the local terminal. But because it does exit with the proper code and passes back a single line of text, it's good enough for Nagios. It's a plug-in that calls another plug-in via Secure Shell, but Nagios doesn't know about this, or care, as long as an exit code and some text are returned.

This methodology is the basis for how any remote execution works in Nagios. Instead of building network protocols into Nagios, the Nagios daemon offloads all that functionality into single purpose plug-ins, which, in turn, communicate to other plug-ins through the protocols of their choosing.

The arrangement outlined above with SSH is not ideal, however. For starters, our remote plug-in has a static threshold. It will always check the load average against 1, so if server2 needs a different threshold because it normally works harder, then server2 also needs a different plug-in. Obviously, this won't scale; we need a way to centrally manage thresholds securely.

The second big problem is SSH authentication. For Nagios to call the remote execution plug-in without being prompted for a password, an authentication key is required. Unless some careful configuration is done, this key could be used to execute anything at all on the remote server, which breaks the principle of least privilege. We want to be able to specify exactly what the Nagios server has access to execute on each host.

These problems, and more, are solved by the Nagios Remote Plugin Executor (NRPE). NRPE has two parts: a plug-in called check_nrpe, which is executed locally by Nagios, and a daemon, which runs on the monitored hosts. The daemon, run via a super server, such as xinetd, or as a service in Windows, has a local configuration file, which defines the commands check_nrpe is allowed to ask for. The check_nrpe plug-in is configured to ask the daemon to execute one of these predefined commands, optionally passing it thresholds. The daemon does so, providing the client output and an exit code, which the client, in turn, passes to Nagios. Any program can be securely executed on the remote server by NRPE, as long as it's defined in the daemon's configuration file. X509 certificates can be used to authenticate the client to the daemon and encrypt the transmission. NRPE is completely cross-platform, so it can handle remote execution for Windows and UNIX clients of all flavors. We'll cover the installation and configuration of NRPE, as well as a few interesting alternatives that might work better for your environment, in Chapter 6, "Watching: Monitoring Through the Nagios Plug-ins."

Scheduling

Now that you have a good understanding of what plug-ins are and how they work, I will explain how Nagios goes about scheduling them. The core of Nagios is a smart scheduler with many user-defined options that allow you to influence the way it goes about its task. Understanding how the scheduler works is imperative to configuring these settings to work with your environment.

Check Interval and States

All internal Nagios processes, including host checks and service checks, are placed in a global event queue. The scheduling of check events is user defined, but not by using absolute date/time in the way cron or the Windows Task Scheduler would. Strictly scheduling dates and times for given services is not possible because Nagios cannot control how long it takes a given plug-in to execute. Instead, you tell Nagios how long to wait after a plug-in has exited before it is executed again. Two options combine to define this time interval.

The interval length defines a block of time in seconds, and the normal check interval[4] is the number of interval lengths to wait. Because the interval length is set to 60 seconds by default, you may think of the normal check interval as the number of minutes to wait between checks.

As depicted in Figure 2.2, events are scheduled by inserting them into the event queue, stamped with the time when they should be run. After Nagios executes a plug-in, it waits for the plug-in to return and then adds the check_interval to the last scheduled run time, to decide when the plug-in should be run next. It's important to stress that Nagios always uses the time the plug-in was originally scheduled to calculate the next execution time, because there are two scenarios where Nagios may need to reschedule a check as a result.

First, if Nagios gets busy and is unable to execute a check at the time it was supposed to, the plug-in's schedule is said to have "slipped." Even if slippage occurs, Nagios uses the initially scheduled time to calculate the next execution time, rather than the time the plug-in was actually run. If the schedule has slipped so badly that the current time is already past the normal check interval, Nagios will reschedule the plug-in.

Second, the plug-in sometimes takes longer to return than expected, due to network delays or high utilization. In the event that the plug-in execution time exceeds the normal check interval, Nagios will reschedule the plug-in.

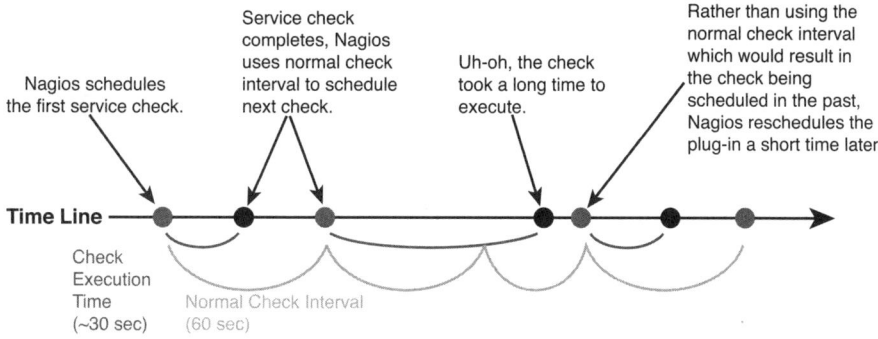

Figure 2.2   Event scheduling

If you thought the term "*normal* check interval" implied the existence of an abnormal check interval, you're on the right track. In the event that a service check returns a code other than 0 (OK), Nagios reschedules the check using a different interval. This setting, called the "retry check interval," is used to do a double take on the service. In fact, Nagios can be configured to do multiple retries of a service, to make absolutely sure the service is down or to ensure the service is down for a certain amount of time, before contacts are notified.

The "max check attempts" option defines how many times Nagios will retry a service. The first "normal" check event counts as an attempt, so a max check attempts setting of 1 will cause Nagios to not retry the service check at all. Figure 2.3 depicts the event timeline for a "max check attempts" setting of 3.

The time between the service first being detected as down and the time Nagios decides to send notifications can be calculated in minutes as:

```
(('retry check interval'*'interval length')*'max check attempts')/60
```

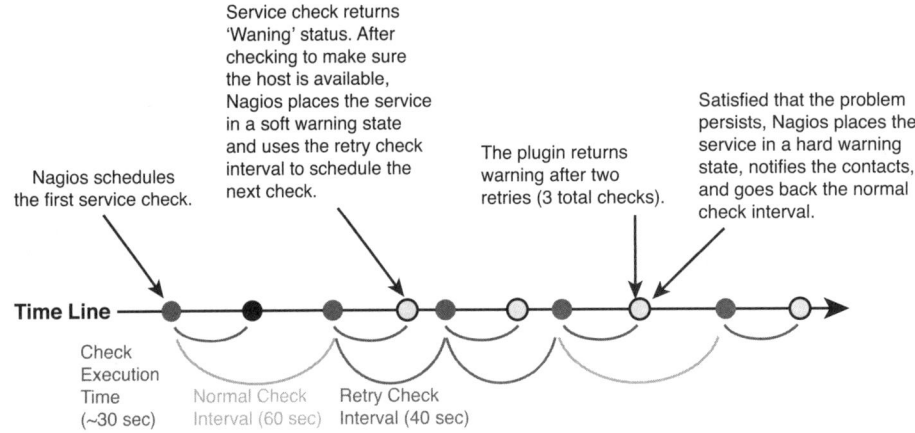

Figure 2.3   Event scheduling during problems

In English, that's the number of retries times the number of minutes to wait between retries. Nagios places the service in a "soft state" while it carries out these retries. After the service is verified as down, Nagios places it in a hard state. Soft states can be one of two types: soft error states, which occur when a service or host changes from okay to something worse, and soft recovery states, when a service or host changes from something bad to something not as bad. Soft states are useful for providing a buffer of time within which to take automatic break-fix steps; because they are logged, they can be useful in detecting services with a tendency to go up and down (or flap) without alerting anyone unnecessarily. The manner in which automatic break-fix is accomplished in Nagios is called an *event handler*. Event handlers are commands that are executed when state changes occur in hosts or services. As usual, the command syntax is, literally, defined by you, so they can do just about anything you want them to do. There are global event handlers, which are executed for every state change, programwide, as well as event handlers you can define on a host-by-host or service-by-service basis. In addition to break-fix, they are a popular place to hook in custom logging or to communicate changes to other monitoring systems.

Distributing the Load

My description of service scheduling in Nagios presents a problem. Because scheduling is determined based on the last time a service completed, the entire scheduling algorithm is dependent on when the services started. If all services started at the same time Nagios did, all services with the same normal check interval would be scheduled at exactly the same time. If this were to happen, Nagios would quickly become unusable in large environments, so Nagios attempts to protect the monitoring server and its clients from heavy loads by distributing the burden as widely as possible within the time constraints provided and across remote hosts. Nagios does this through a combination of intelligent scheduling methodologies, such as service interleaving and inter-check delay.

When Nagios first starts, it usually does so with a long list of hosts and services. Nagios's job is to establish the status of each element on that list as quickly as possible so, in theory, it could just go down the list item by item until it came to the bottom and then begin again from the top. This methodology is not optimal, however, because working the list from top to bottom puts a lot of load on individual remote hosts. If server1, at the top of the list, were configured with 18 services, for example, Nagios would demand the status of all 18 immediately. Instead, Nagios uses an interleave factor, as depicted in Figure 2.4.

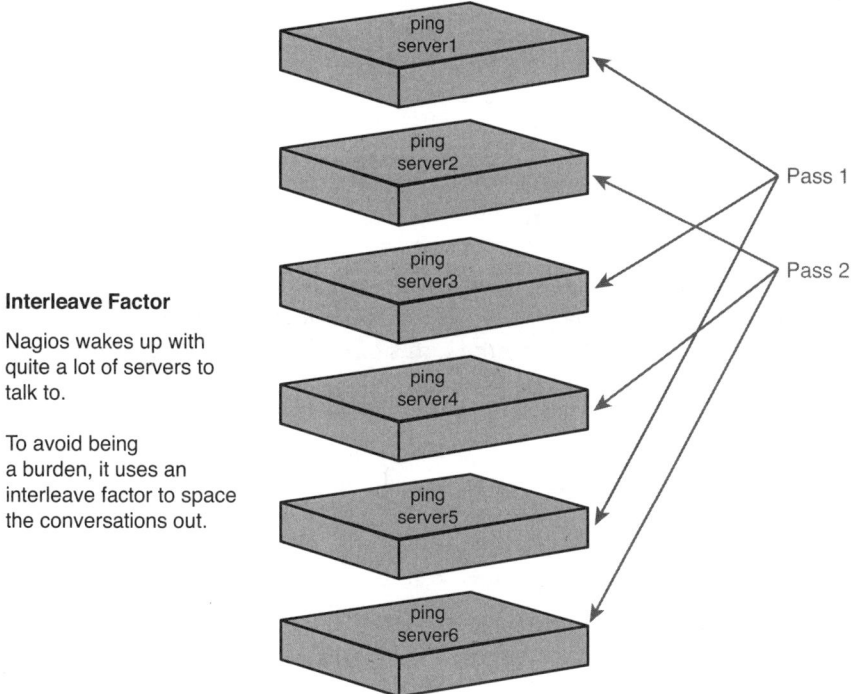

Interleave Factor

Nagios wakes up with quite a lot of servers to talk to.

To avoid being a burden, it uses an interleave factor to space the conversations out.

Figure 2.4 With an interleave factor of 3, Nagios checks every third service.

With an interleave factor of 2, Nagios would ask for every other item in the list until it got to the bottom: 1, 3, 5, and so on. Then, beginning from the second to the top, it would ask for every other again: 2, 4, 6, and so on. In this manner, the load is distributed across different servers, yet the overall state of the network takes no longer to discover than with the top-down method. The interleave factor is user definable, but Nagios, by default, will calculate one for you using the following formula, which is optimal in nearly all cases:

```
interleave factor = ( total number of services / total number of
hosts )
```

Reducing the load on the remote hosts, however, does nothing for the Nagios server, which still needs to process the sending and receiving of large numbers of checks upon startup. To alleviate the intensive burden of the first few minutes after startup and to ensure that cycles are available for important tasks, such as log rotation, Nagios inserts a small amount of time between checks that would otherwise be executed in parallel. This time period, called the inter-check delay, is user-definable, but if set too high, may cause the schedule to slip. Nagios will calculate one for you if the inter-check delay method is set to smart (which it is, by default) using the following formula:

```
inter-check delay = (average check interval for all services) /
(total number of services)
```

You should be aware that many other options exist to influence the scheduler, especially during startup. For example, a max check spread can be imposed, which limits the amount of time Nagios may take to initially glean the status of each host. Obviously, this option, if set, affects the previous inter-check delay formula. To help you sort out some of these settings, execute the Nagios binary with an '-s switch to provide scheduling information based on the current configuration. Nagios XI also has some very nice built-in visualization of the scheduler state, which can make diagnosing scheduling problems much easier.

Reapers and Parallel Execution

The events that are processed in the queue break down into one of two types: those that can be run in parallel and those that cannot. Service checks in Nagios can be run in parallel, so when a service check event is found in the queue, it is forked, and Nagios continues on to process the next event in the queue. The service check that was forked will execute a plug-in and place its error code and output in a message queue until an event called a "reaper" comes along to collect it.

Reaper events are the heartbeat of Nagios. Their frequency of occurrence in the event queue is a user-defined option, and it's an important one. No matter how fast the plug-in finishes its work and reports back its status, Nagios will not process it until a service reaper comes along and discovers it in the message queue.

The number of service events Nagios is allowed to execute in parallel is defined by the max concurrent checks option. This variable is also an important one; if set too high, the monitoring server's resources will be completely consumed by service checks. On the other hand, set it too low, and service check schedules may slip. The Nagios documentation provides a primer about how to optimize the concurrent checks variable, based on the average amount of time it takes for your plug-ins to execute, here:

```
http://nagios.sourceforge.net/docs/3_0/checkscheduling.html#max_
concurrent_checks
```

The Nagios web interface includes tools to quantify the number of checks that are not meeting their schedules. In practice, users are usually needlessly concerned. Some number of events will inevitably miss their initially scheduled windows because they've been preempted by retry checks for other services that have gone into soft error states, because some checks take longer than others, or a plethora of other reasons. So don't worry too much if you see some schedules slip now and again; the Nagios scheduler, in my experience, generally does the right thing, given the chaotic nature of its task.

Notification

If you haven't configured it to do otherwise, Nagios will send notifications every time a hard state change occurs[5] or whenever a host or service stays in a bad state long enough for a follow-up notification to be sent. So many options in various object definitions can affect whether notifications are sent that it can be difficult to maintain a mental picture of the overall notification logic. Nagios's notification framework is a welcome surprise to most sysadmins for the robust flexibility it provides, but it is also an often-misunderstood subject among neophytes because it is so configurable.

I've found that describing the notification logic is a great way to introduce a lot of fundamental concepts. If you can understand how notifications work within Nagios, you understand a great deal about Nagios. I start from the top and explain the various levels for which notification options can be configured and point out a few potential "gotchas" along the way.

Global Gotchas

Nagios, like countless Unix programs before it, is configured by way of text files. So generally, the definitions in the text files will determine Nagios's state while it's running. However, certain settings can be changed at runtime, so the config file definitions don't necessarily reflect what Nagios is doing for a few specific settings. One important example is the global enable notifications setting. This option enables or disables notifications programwide and it's probably the most dangerous of the runtime changeable configuration options.

It is especially important to be aware of the actual state of enable notifications because Nagios can be configured to use a persistent program state. That is, when Nagios is shut down, it writes its currently running configuration state to a file, so that when it is started again, it starts right where it left off. When Nagios is started in a persistent program state, via the use retained program option, the settings in the state file override those in the configuration files. This means that if the enable notifications option is set to enable in the config file, but then is reset to disable while Nagios is running, it's possible that this setting will persist across application restarts, and someone looking in the configs will believe that it is, in fact, enabled, when it is not.

Always make sure to check the "Tactical Overview" CGI from the Nagios GUI if you have notification problems, to make sure that notifications are globally enabled. If you feel that restarting Nagios may resolve your problem, be sure to disable program state persistence and/or delete the state file before you start Nagios back up. Deleting the state file while Nagios is running is no good; Nagios will simply write a new one when it is shut down. So shut down first, then delete, and then start back up.

Notification Options

Hosts and services each have several options that affect the behavior of notifications in Nagios. The first to be aware of is the notification options setting. Each host or service can be configured to send notifications (or not) for every possible state that Nagios tracks. These states are slightly different for hosts and services because hosts and services are different in reality. Table 2.2 summarizes the possible host and service states.

Table 2.2   *Host and Service Notification States*

Host States	Service States
Unreachable (u)	Unknown (u)
Down (d)	Critical (c)
Recovered (r)	Warning (w)
Flapping (f)	Recovered (r)
	Flapping (f)

"Flapping" is the term Nagios coined to describe services that go up and down repeatedly. Nagios can be configured to detect flapping services as a notification convenience, because service flapping can cause many unwanted notifications. The notification options setting should list each state for which you want Nagios to send a notification. If critical CPU notifications are desired, but not warnings, the notification options setting should list only c. If you want to be notified when a service recovers from a bad state, then r must be explicitly listed, as well.

In addition to service and host definitions, contact definitions also have a notification options setting. Contact definitions are used to describe a person to whom Nagios may send a notification. In addition to obvious options, such as the contact's name and email address, each contact definition may contain a host notification options setting and a service notification options setting. Together these options provide the capability to filter the type of notifications any single contact receives. A programmer, for example, might always want problem notifications for the applications he's responsible for, but never want recovery pages because he knows when the problem is fixed by virtue of the fact that he's the one fixing it.

Templates

A potential gotcha, with definitions in general, is the concept of definition templates. Because there is a lot to define, Nagios allows you to create generic definitions that list options that certain types of systems have in common. The web servers may all have common notification contacts and thresholds, for example, so when you go about creating the service definitions for the web servers, you might create a generic service definition called web-servers first, and refer the individual service definitions to it. This way, you have to define thresholds and notification contacts only once, instead of once per service. These generic definitions are called templates, and they save you a substantial amount of typing.

Options explicitly defined in a service definition take precedence over those set in the template that the definition refers to. However, when the definition inherits its notification options setting from a template, it isn't immediately obvious what states the service is set to notify on. I always try to explicitly set the notification options, and I recommend that you do, too. It's common for people to cut and paste a lot when dealing with Nagios configs, so at least be aware of the notification options setting and know which template it inherits from, if it's not explicitly set. More information on templates can be found in Chapter 4, "Configuring Nagios."

Time Periods

Something else that will affect Nagios's notification behavior is the configuration setting for time periods. Like many other things in Nagios, time periods are user defined. They are used to specify a block of time, such as Monday through Friday, 9 a.m. to 6 p.m., which service and host definitions refer to in order to derive their hours of operation. Service definitions refer to time periods in two ways. The notification period defines the hours within which

Nagios is allowed to send notifications for the service, and the check period defines the period of time Nagios may schedule checks of the service.

Nagios isn't just a monitoring system; it's also an information collection program. Nagios can collect utilization and status information from anything it monitors and pass this information along to other programs for graphing or data mining. Services such as CPU utilization are good candidates for 24 x 7 data collection, but you may not want to send notifications all the time, because CPU-intensive things, such as backups, tend to happen at night. This is a perfect example of why you might want to have different settings for notification period and check period.

If the service breaches its thresholds outside of *either* of these time periods, Nagios will not send a notification. If the threshold breach occurs outside of the notification period, Nagios will track the state change from OK into Soft Error and into Hard Error, but it will not send notifications because it has been explicitly told not to do so. Alternatively, if the threshold is breached outside the check period, Nagios will not notice, because it has been explicitly told not to schedule checks of the service. Like almost every other option within the service definition, the time periods may be inherited from a template, if not explicitly set, so be aware of what time periods are being inherited if you have notification trouble.

As with the notification options setting, time periods may be configured on a contact-by-contact basis, so even if a threshold breach occurs within the time period specified by the service definition, the contact might still filter out notifications.

Scheduled Downtime, Acknowledgments, and Escalations

The last few variables that could affect Nagios's notification decision are escalations, acknowledgments, and scheduled downtime. In the event that planned maintenance must take place on one or more hosts, it is possible to schedule a period of downtime for the host or hosts from the Nagios web UI. While a host is in a period of scheduled downtime, Nagios will schedule checks as normal and continue to track the host's state, but no notifications will be sent for that host or its services. Further, Nagios distinguishes between actual downtime and scheduled downtime in its various web reports.

Escalations exist as a means to notify additional contacts if a host or service is down for too long. For example, if server1 goes down, you may want Nagios to notify the sysadmin, but if server1 is down for 7 hours, you may want to notify the sysadmin and his manager. Escalations are defined independently, so they are not part of the service definition itself. When Nagios decides the time is right to send a notification, it first checks to make sure there isn't an escalation definition that matches the notification Nagios is about to send. If Nagios finds a matching escalation, it will send it instead of the original notification.

It's important to realize that Nagios will *either* send the notification as defined in the service *or* the notification defined in the escalation. Normally, this is fine because the escalation and the notification in the service are the same thing, except the escalation has a few extra contacts. However, it is possible to define the escalation with a completely different set of contacts and even a completely different notification command. Be sure to list the original contacts in the escalation definition, along with the upper-tier contacts, if you intend for both the sysadmin *and* his manager to be notified.

Acknowledgments are used to silence the reoccurring follow-up pages while you work on fixing the problem. In the event that escalations are configured, it's especially important to tell Nagios that you're aware of and working on the problem, so it doesn't go and get your manager involved. Acknowledgments can be sent to Nagios by way of the web interface, along with an optional comment. When an acknowledgment is sent, Nagios notifies the original recipient list that the problem has been acknowledged and by whom. Follow-up notifications are then disabled until the service goes back into an OK state.

I/O Interfaces Summarized

Nagios is a great monitoring tool, but the area in which it is head and shoulders above any of its commercial brethren is its capability to interact with other external monitoring and visualization tools. Nagios is good at interacting with other systems because this was a design goal of its creators, but also because, as we've repeatedly stated in this chapter, it has little functionality built into it besides scheduling and notification. Nagios is good at making it easy to get data in and out because it *has* to be good at it, by virtue of its simple yet elegant design. In this section, I describe a few of the most common ways to get data into and out of Nagios.

The Web Interface

When you build Nagios, you may also choose to build its web interface, and I recommend that you do. The main purpose of the web GUI is to provide you with a window into the current state of the hosts you are monitoring, but it has a lot of functionality beyond that, such as the following:

- Historical reporting and trending tools
- Interfaces for scheduling downtime and providing comments to specific hosts and services
- Interfaces for enabling or disabling service checks and notifications
- Interfaces for examining the current configuration
- Tools for drawing graphical maps of the environment
- Tools for getting information about the Nagios daemon's status

The web GUI is CGI based, and the CGI programs are written in C, so they run fast. In general, the CGIs glean their information from several log and state files that Nagios maintains in its var directory. The complete functionality of the GUI is a subject that could fill an entire book, so I'm going to summarize its elements and how they tie together, and I'll also show you some of my personal favorite displays; the ones I find myself revisiting often. With the top-level summary that follows, you should be able to explore the GUI on your own and glean the information you need in no time.

Figure 2.5 depicts the Nagios Navigation bar. The bar is organized into four sections: General, Monitoring, Reporting, and Configuration. The General section has a link to the full Nagios documentation in HTML. The Configuration section can be used to look at online versions of the configuration files. The two more interesting sections are Monitoring and Reporting.

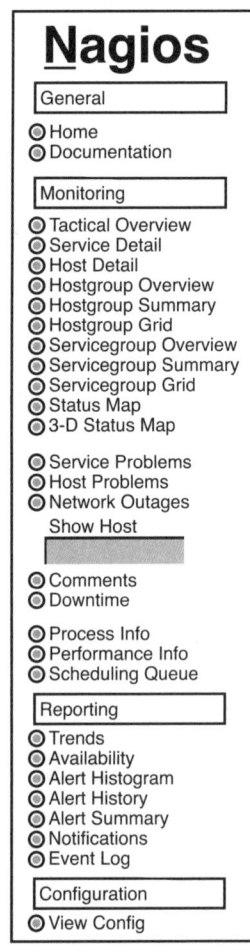

Figure 2.5   The Nagios Navigation bar

Monitoring

The Monitoring section is intended to provide real-time status displays of the machines you monitor with Nagios. With the exception of the tactical overview, the status map, and the 3D status map, every display available under the monitoring section is provided by a single CGI: status.cgi. In fact, 90 percent of what you will probably do with the web interface is interaction with status.cgi, and this is a good thing because status.cgi provides a uniform interface. In general, every screen displayed by status.cgi will have four elements in common across the top of the display: a context-sensitive menu of links, the host status table, the service status table, and the display area. These are shown in Figure 2.6.

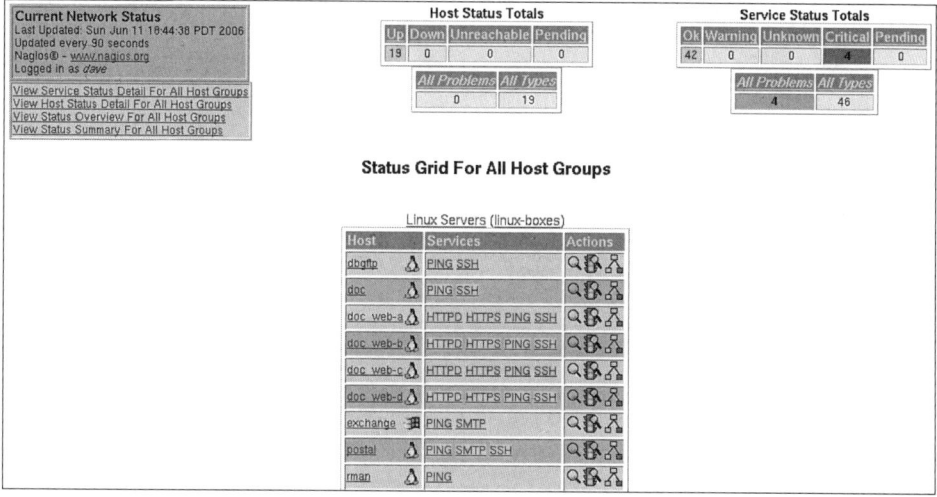

Figure 2.6   The status.cgi display

The link menu in the upper left provides shortcut links, which change based on what type of display you look at, and the service and host status tables on the upper right summarize the state of the hosts and services in the display area. It's important to note that the host and service status tables are context sensitive; that is, they display a summary of the subset of hosts and services you are currently viewing in the display area and *not* the overall status of every host and service in Nagios. Generally, clicking links in the display area gives you more detailed views. Clicking a hostgroup gets you a summary display of all the hosts in that group. Clicking a host gets you the detail display of that host.

Figure 2.7 is a screenshot of the detail display for a host. As you can see, the host and service status displays are gone, and the display area contains a table listing the current status of the host, as well as a menu of commands. These commands alter the status of this host. From here, you can schedule downtime for the host, tell Nagios to stop checking it, stop sending notifications about it, acknowledge problems, and so on. If Nagios is configured to use a persistent state, via the retained state option, these changes will persist even after Nagios is restarted.

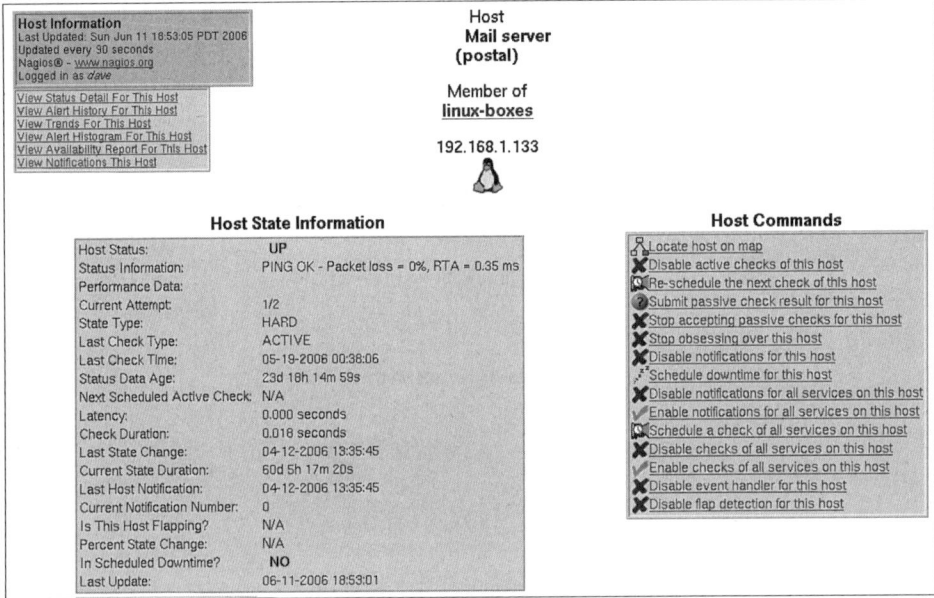

Figure 2.7   Status detail display for a host

Figure 2.8 is a screenshot of the hostgroup summary display. This is my favorite display; it shows, on a single screen, the state of the entire monitored environment, broken down by hostgroup. I like it mostly because, no matter how large your environment, you can usually fit the entire summary screen within a 1024 x 768 window. The hostgroup summary is where I go in the morning to see the status of the environment at a glance and it's also the screen I keep up to glance at throughout the day. From here, you can drill down into any hostgroup, and on down to the host, and because it includes every host and service in the entire environment, the host and service status tables show the current status of the environment as a whole.

Reporting

Nagios has some excellent reporting capabilities built in. The Reporting CGIs break down into one of three types: those that just provide log dumps, those that draw graphs, and those that output formatted reports. Alert History, Notifications, and Event Log simply dump the log lines that correspond to their type. Alert Histogram and Trends use the GD library to generate graphs. Availability and Alert Summary generate formatted reports.

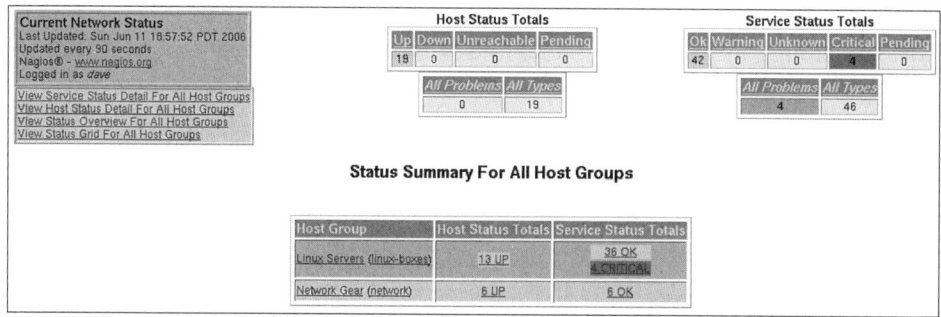

Figure 2.8   Hostgroup summary: my favorite screen

All the reporting types that aren't just log dumpers follow the same pregeneration procedure. Click the report type and you will be prompted to specify which type of object you want to run the report on. You may run most reports on one or more hosts, services, hostgroups, and servicegroups. After you choose the type of object, you are prompted to select the specific object you want to run the report on, so if you choose host(s) in the first step, you can specify server1 in the second. Last, you are asked to provide some options, such as the report time period and which state types (hard or soft, error or non-error) you want it to include.

The main difference between a trend graph and a histogram graph is the X-axis. For a period of five days, a trend graph displays each of those five days on the X-axis and plots the service state (unknown, critical, warning, okay) on Y. The trend graph is a straightforward historical representation of the state of the service as a function of time.

The histogram, on the other hand, displays a user-defined breakdown of time intervals (referred to as the breakdown type) on the X-axis, such as hours per day or days per week. It then plots the number of error states on the Y-axis. For example, a histogram with a period of Last 7 days and a breakdown type of hours per day will plot 24 hours on the X-axis and then plot the number of occurrences of each problem state in the last seven days on the Y-axis, based on the hour it occurred. This is handy for visualizing trends in service outages A trend graph will show you that the CPU was High on server1 last Monday at 1 p.m., but the histogram will show you that the CPU is high on server1 *every* Monday at 1 p.m.

Both the Availability and Alert Summary formatted reports are well done, and I use them often. The availability report is what you look for if you need to prove you're meeting your SLA. Availability reports are color-coded tables that specify the exact amount of availability (as a percentage to three sig-figs for you "five nines" people) for any host, service, or collection thereof. The Alert Summary is great for people who don't want to get paged, but still want a summary of what went on in the past 24 hours or any other user-defined time period. The Alert Summary can be filtered to exclude soft events, show only host events or service events, or both, or display only problem states, only recovery states, and so on. I've found that both the Availability Report and Alert Summary Report can be easily imported into Excel spreadsheets for manager consumption by saving the report straight out of the HTML frame as whatever.xls.

The External Command File

The fact that the web CGIs are able to do things, such as schedule downtime for hosts and turn off notifications, implies that there is more to the web interface than a simple CGI wrapper to the log and state files. In fact, the CGIs accomplish command execution by way of the external command file. The command file is a FIFO (or named pipe), which Nagios periodically checks for commands. It can be thought of as a file system entry point to the event queue. The command file isn't just for the web interface; it can be used by any process with the permissions to do so.

The format for commands is the following:

```
[time] command_id;command_arguments
```

The Time is a timestamp in Unix seconds since epoch format. To get the current time in seconds on a Unix box, do this:

```
date '+%s'
```

The command ID is the name of the command you want to execute and each one takes different arguments. As of this writing, there are 131 different commands you can send to the Nagios daemon, by way of the command file, to do things such as turn on and off notifications, acknowledge problems, and provide comments. A listing of these commands, along with their options and sample shell scripts for each one, is available from the following:

```
http://nagios.sourceforge.net/docs/nagioscore/3/en/extcommands.html
```

Performance Data

In the section "Plug-ins," we defined a plug-in as a little program that provides an exit code and optionally some text on stdout. If the optional text contains a pipe character (|), Nagios will treat anything before the pipe as normal summary text and everything after the pipe as performance data. Nagios can be configured to handle performance data differently than other types of data if it exists.

In practice, performance data is the de facto means to export data from Nagios into graphing programs, such as RRDTool. Most plug-ins, however, do not provide performance data; that is, they have no pipe character in their summary text. This is easy enough to rectify, however, with the check_ wrapper_generic in Listing 2.6.

Listing 2.6   *A Performance Data Wrapper for All Plug-ins*

```
#!/bin/sh
#a wrapper which adds perfdata functionality to any nagios plugin
#link pluginName_wrapper to this script for it to work
#for example, if you want to enable perfdata for check_mem
#you would 'ln -s check_wrapper_generic check_mem_wrapper'

#get rid of the 'wrapper' on the end of the name
NAME='echo $0 | sed -e 's/_wrapper//''

#call the plugin and capture its output
OUTPUT='${NAME} $@'
#capture its return code too
CODE=$?

#parrot the plugin's output back to stdout twice, separated with a
pipe
echo "${OUTPUT}|${OUTPUT}"

#exit with the same code that plugin would have exited with
exit ${CODE}
```

This wrapper script is similar to the SSH remote execution plug-in in Listing 2.5. It calls another plug-in by proxy and parrots back the other plug-in's output, this time adding a pipe in the output, thereby adding support for performance data. In Chapter 8, "Visualization." I talk more about performance data and how it can be used to feed visualization programs.

The Event Broker

The Event Broker is a process that watches the event queue for certain types of events to occur and then notifies one or more Event Broker modules, passing relevant details about the event to the module.

The NEB modules are written in C and linked to the Nagios core process at runtime, at which point they ask the event broker to send them the events they are interested in. These events can be any type that Nagios deals with in the event queue. After an interesting event occurs and the module has the relevant details, it may do almost anything it wants, including modifying events in the queue, passing information to other programs, and generally changing the way Nagios operates.

The Event Broker interface is absolutely the most powerful interface to Nagios, and several interesting Event Broker Modules exist to extend Nagios functionality. Several of these I cover in Chapter 7, "Scaling Nagios." For those with some C skills and the inclination to get their hands dirty, I've included Chapter 10, "The Nagios Event Broker Interface," where I walk you through creating an NEB module by using the event broker to add a file system status interface to Nagios.

End Notes

[1] There may be multiple MXs.

[2] http://nagiosplug.sourceforge.net

[3] Every programming and scripting language that I'm aware of can provide exit codes.

[4] It should be noted that normal check interval applies only to service definitions. Although it is possible to specify a check interval for hosts, it is not required for Nagios to work properly. In general, host checks are carried out as they are needed, usually after a service check on that host fails, so the use of explicit host checks is discouraged.

[5] See the section "Check Interval and States" for a description of hard states.

## CHAPTER 3

Installing Nagios

Nagios is composed of three chunks: the daemon, the Web interface, and the plug-ins. The daemon and Web interface make up the tarball available from www.nagios.com, and the plug-ins must be downloaded and installed separately.[1] Although the Nagios daemon comes with the Web interface, the daemon may be installed alone. Nagios does not require the Web interface to effectively monitor the environment and to send notifications.

After the Daemon is installed, most people download and install the plug-ins tarball. Installing the plug-ins is technically an optional step; it's entirely possible to run Nagios with nothing but custom plug-ins, but most people consider that to be a reinvention of the wheel. The plug-ins provided by the Nagios Plugin Project are well written and contain most of what you need to monitor an enterprise network. This chapter covers the installation of all three chunks and provides a handy reference to the available configuration options.

OS Support and the FHS

The Nagios daemon was designed for and on Linux, but it is capable of being run by any UNIX-ish operating system, including Solaris, BSD, AIX, and even Mac OS X. There are rumors on the mailing lists of success running the daemon under Cygwin on Microsoft Windows, but I haven't personally seen it done.[2] Fair warning: I happen to run Nagios on Linux, so what I write may have a bit of a Linux slant, although I will try to keep it to a minimum.

The main difference between various UNIX environments running Nagios is the file system hierarchy standards associated with each. Unfortunately, different UNIX variants will place the same file in different locations, which makes it difficult to predict where particular files might end up. It's a problem that dates back to the BSD split, and no one has come up with a good solution (or perhaps everyone has and they're all different). Even between different distributions of Linux, differences may exist in the file system hierarchy implementation. The only real way to know where all the files will wind up is if you manually install from source, or use a source-based distro.

Aside from some odd constructs, such as the AIX convention of installing everything open source into "/opt/freeware," most systems will install Nagios in one of two ways: using either FHS or installing into "/usr/local." The File System Hierarchy Standard (FHS) is a Free Standards Group proposed standard and describes where files should go in UNIX-like file systems. Most binary Linux distributions I'm aware of, such as Red Hat, Mandrivia, and SuSE, as well as some source-based distributions such as Sourcemage GNU Linux, will use the FHS to some degree. I'd like to emphasize that, although the FHS is a good standard, there is still a lot of disagreement and confusion as to how it works in practice. So Table 3.1, which depicts the FHS Standard file locations for Nagios, should be considered a rough guide.

Table 3.1   *Nagios File Locations in the FHS*

File Type	Location
Configuration Files	/etc/nagios
HTML	/usr/share/nagios
CGIs	/usr/share/nagios or /usr/lib/nagios
Program daemon and other executables	/usr/bin/nagios
LockFiles and FIFOs	/var/lib/nagios or /var/log/nagios
Logs	/var/log/nagios/
Plug-ins	/usr/libexec/nagios or /usr/lib/nagios

If you use a source-based distribution such as Gentoo or manually install Nagios from source on Linux, Solaris, or the BSDs, expect to find everything under "/usr/local/nagios" or "/usr/nagios," unless you specify different locations with options to the configure script. Technically, "/usr/local/nagios" is consistent with the FHS, which states that locally installed software—that is, software not installed via the system's package manager—should be installed to "/usr/local." Table 3.2 lists the file locations for a "local" installation.

Table 3.2 *Nagios File Locations for Local Installs*

File Type	Location
Configuration Files	/usr/local/nagios/etc
HTML	/usr/local/nagios/share
CGIs	/usr/local/nagios/share
Program daemon and other executables	/usr/local/nagios/bin
LockFiles and FIFOs	/usr/local/nagios/var
Logs	/usr/local/nagios/var
Plug-ins	/usr/local/nagios/libexec

Installation Steps and Prerequisites

Let's get this show on the road. The high-level steps to install Nagios are as follows:

1. Obtain and install the Nagios dependencies.
2. Obtain and install Nagios.
3. Obtain and install the plug-ins' dependencies.
4. Obtain and install the plug-ins.

Nagios has only a few dependencies, and most of them are optional. If you want to use the Web front-end, which is the only interactive interface available, you need a Web server with CGI support, such as Apache.[3] Three graphics libraries are needed if you want Nagios to display pretty pictures and graphs: libpng, libjpeg, and the gd library. The only required dependency is zlib, and chances are you already have that.

The plug-ins' dependencies are more of a moving target. You will need a ping program, some BIND tools, such as host, dig, or nslookup, the OpenSSL library, and Perl. If you plan on querying network objects with SNMP, you need net-snmp and possibly perl-snmp. Depending on what you need to monitor, packages such as OpenLDAP, Kerberos, and MySQL/pgsql may be needed for special purpose plug-ins.

On most OSs, the configure script in the plug-ins package is good at figuring out what packages you have and automatically building plug-ins for them. If you lack a package, such as OpenLDAP, configure will not attempt to build the ldap plug-ins, rather than generate an error.[4] The base directory of the plug-ins' tarball contains a REQUIREMENTS file, which lists the requirements of specific plug-ins, so make sure to check it out before you build.

Installing Nagios

Most UNIX operating systems I'm aware of ship with prepackaged versions of Nagios or otherwise have them available. In the Linux realm, Red Hat, SuSE, and Mandrivia users will find Nagios RPMs on their installation media,[5] Debian and Ubuntu types may "apt-get install nagios-text," gentoo pundits can "emerge nagios," and even sourcemage ... wizards(?) may "cast nagios." BSD people will find Nagios in ports and Solaris folks can "pkg-get install Nagios" from blastwave.org. No packages currently exist for AIX.[6] Consult your local package manager documentation for more information on installing the Nagios package available for your system.

For manual builds from source, the Nagios tarball may be obtained directly from www.nagios.com. The installation process is a typical configure, make, and make install. Nagios requires a user and group to run, so these must be created first. For the impatient person, Listing 3.1 should be enough to get you up and running.

Listing 3.1   *Installing Nagios for the Impatient Person*

```
groupadd nagios
useradd -s /bin/false -g nagios nagios
tar -zxvf nagios-version.tgz
cd nagios-version
./configure
make all
sudo make install
```

As with many great open source applications, configure and build options abound. Let's take a closer look at the three main steps: configure, make, and make install.

Configure

The defaults are sensible and you aren't required to specify anything, but you should be aware of a few options. Launch configure with a '-h' switch for a full list of options. In my experience, you either need to change nearly all of them or few to none. Table 3.3 shows some options you should be aware of upfront.

Table 3.3   *Important Compile-Time Options*

--enable-embedded-perl	This enables Nagios' embedded Perl interpreter. It is intended to speed up execution of Perl scripts for people who use a lot of Perl-based plug-ins. I bring it up because it has a reputation for causing segfaults, so turn it on at your own risk. The default is 'disabled' and most people keep it that way.
--with-nagios-user=<usr>	This is the user that Nagios runs as. It should be created before running configure. The default is 'nagios.'
--with-nagios-group=<grp>	This is the group that the Nagios user belongs to. It should be created before running configure. The default is 'nagios.'
--with-command-user=<usr>	The username used to segment access to the Nagios command file, which is described at the end of Chapter 2, "Theory of Operations." It defaults to the 'nagios-user' option.
--with-command-group=<grp>	This specifies the group used to segment access to the Nagios command file, which is described at the end of Chapter 2. It defaults to the 'nagios-group' option.
--with-init-dir=<path>	This sets the location that the init script will be installed in, if you use the 'install-init' make target. The default is '/etc/rc.d/'.
--with-htmurl=<path>	The CGIs that make up the Web front-end are written in C, so the URL paths in the HTML they generate must be set at compile time. This defaults to '/nagios/,' meaning that the HTML generated by the CGIs will reference 'http://localhost/nagios/'.
--with-cgiurl=<path>	Similarly, any CGI URL paths referenced by the HTML that the Web front-end generates must be set at compile time. This defaults to '/nagios/cgi-bin/,' meaning that the HTML generated by the CGIs will reference 'http://localhost/nagios/cgi-bin/'.

The previous nine options are good litmus tests for whether you will have to specify options to configure. A "./configure" with no options at all will get most people a working Nagios implementation, but if you need to change more than two of the preceding defaults, chances are you'll have to change something that I haven't listed.

Make

There are, in fact, five make targets to build various pieces of Nagios. A simple 'make all' will build Nagios and the Web front-end. Table 3.4 lists the other build-related make targets.

Table 3.4   *Build-Related Make Targets*

make nagios	Just make the Nagios Daemon, without the Web interface.
make cgis	Make only the Web interface, without the daemon.
make modules	Make the Event Broker Modules. As of this writing, there's only one module included: helloworld.o, which is really useful only for people who want to learn how to program NEB Modules.
make contrib	Make various user-contributed programs. There are five of these. The more useful ones include a traceroute CGI, a sample apache config file, and a standalone version of the embedded Perl interpreter for testing Perl plug-ins.

Make Install

Executing `make install` will install the daemon, CGIs, and HTML files. Eight other install-related make targets exist, as defined in Table 3.5.

Table 3.5   *Install-Related Make Targets*

make install-base	Install only the Nagios daemon without the Web front-end.
make install-cgis	Install the CGI programs.
make install-html	Install only the static HTML pages.
make install-init	Install the init file (to the directory specified by the –with-init-dir configure option).
make install-config	Install sample configuration files.
make install-commandmode	Create the external command file (as described at the end of Chapter 2).
make uninstall	Uninstall Nagios.
make fullinstall	Install the kitchen sink (everything).

Putting it all together, a "real" Nagios build may look more like Listing 3.2.

Listing 3.2   *A Realistic Nagios Installation*

```
groupadd nagios
useradd -s /bin/false -g nagios nagios
useradd -s /bin/false -g nagios nagioscmd
tar -zxvf nagios-version.tgz
cd nagios-version
./configure -with-command-user=nagioscmd
make all
sudo make install install-init install-config install-commandmode
```

After `configure` finishes running, it provides you with a handy summary of what happens if you decide to build. For example, running 'configure' on my Linux workstation with the options in Listing 3.2 gives the summary in Listing 3.3.

Listing 3.3 *Output from 'configure'*

```
*** Configuration summary for nagios 2.1 03-27-2006 ***:

General Options:
-------------------------
       Nagios executable:  nagios
      Nagios user/group:  nagios,nagios
     Command user/group:  nagioscmd,nagios
          Embedded Perl:  no
           Event Broker:  yes
      Install ${prefix}:  /usr/local/nagios
              Lock file:  ${prefix}/var/nagios.lock
         Init directory:  /etc/rc.d
                Host OS:  linux-gnu

Web Interface Options:
----------------------
               HTML URL:  http://localhost/nagios/
                CGI URL:  http://localhost/nagios/cgi-bin/
   Traceroute (used by WAP):  /usr/sbin/traceroute


Review the options above for accuracy. If they look okay,
type 'make all' to compile the main program and CGIs.
```

Installing the Plug-ins

After Nagios is installed, it's time to install the plug-ins tarball so that Nagios can run some checks. The plug-ins tarball and RPMs are available from the downloads section of www.nagios.com. Typically, UNIX distributions that have a Nagios package also have a Nagios plug-ins package of some description.

Manual installation from source code of the plug-ins is easier than Nagios itself. The configure script figures out the paths to important binaries, such as ping and Perl. In the event it can't find something, you may have to specify the location, but this is unlikely. If you specified custom options to Nagios, you may also have to specify them to the plug-ins. These include any of the default install directories you may have changed, as well as those listed in Table 3.6.

Table 3.6   *Configure Options for the Nagios Plug-ins*

--with-cgiurl=<path>	If you specified a custom cgiurl in the Nagios build, you need to tell the plug-ins about it here.
--with-nagios-user=<user>	If you are running Nagios with a nonstandard username, specify it to the configure script for the plug-ins.
--with-nagios-group=<group>	Likewise, if you changed the group away from the default during the Nagios build, change it for the plug-ins as well.
--with-trusted-path=<colon:delimited:list:of:paths>	This very cool option lets you specify a custom PATH for the environment the plug-ins run in. This increases the security of the system by limiting where the plug-ins may go to execute other programs.

Call 'configure' with a '-h' to get a full list of options if, for example, you need to specify the location of the ping command because configure was unable to find it. For most environments, configure can be run with default settings. Similar to the main program, the plug-ins 'configure' script will generate a handy summary like the one in Listing 3.4, which was the result of calling 'configure' on my Linux workstation with no options specified.

Listing 3.4   *Output from Plug-ins 'configure'*

```
              --with-perl: /usr/bin/perl
            --with-cgiurl: /nagios/cgi-bin
       --with-nagios-user: nagios
      --with-nagios-group: nagios
     --with-trusted-path: /bin:/sbin:/usr/bin:/usr/sbin
      --with-ping-command: /bin/ping -n -c %d %s
     --with-ping6-command:
             --with-lwres: no
              --with-ipv6: yes
           --with-openssl: yes
--enable-emulate-getaddrinfo: no
```

If the summary looks good, a simple 'make && make install' will build and install the plugins to the appropriate place. Not installed, however, are the contents of the 'contrib' directory in the base directory of the plug-ins tarball. The contrib directory is a gold mine of special purpose and architecture specific plug-ins. It contains checks for everything from netapp appliances to Sybase databases and everything in between. I highly recommend that you take a look in contrib before developing anything on your own, even if you don't find what you're looking for; it's a great place to go for code to repurpose.

Installing NRPE

After Nagios and the plug-in tarball are installed, you probably want to skip ahead to Chapter 4, "Configuring Nagios," and get them configured. After you have a fully functional Nagios server, however, the next step is remote execution. As described in the section "Scheduling," at the end of Chapter 2, "Theory of Operations," the Nagios Remote Plugin Executor (NRPE) provides Nagios with the capability to execute plug-ins located remotely on the monitored hosts. As depicted in Figure 3.1, NRPE consists of two pieces: a plug-in, which resides on the Nagios server, and a daemon, which runs remotely on each monitored host. Nagios uses the check_nrpe plug-in to ask the NRPE daemon to run a check on the remote host. If NRPE on the remote host is configured to allow this, it runs the plug-in and passes the results back to check_nrpe on the Nagios server.

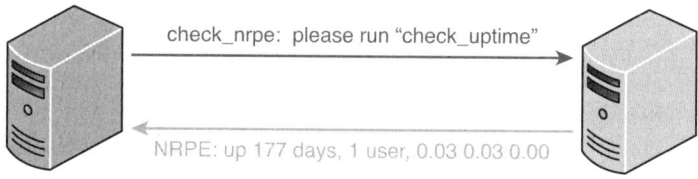

Figure 3.1   Remote execution with NRPE

The NRPE daemon can run under a superserver, such as inetd/xinetd, or it can be run as a standalone daemon. It is configured by way of a config file, called nrpe.cfg, which is usually located in /etc/. The nrpe.cfg defines a list of the plug-ins that the check_nrpe client is allowed to request. There is also a version of the NRPE daemon available for Microsoft Windows. Check Chapter 6, "Watching: Monitoring Through the Nagios Plug-ins," for more information about how to use NRPE to check services on remote hosts.

To install the Linux version of the daemon, first obtain it from the following:

```
http://exchange.nagios.org/directory/Addons/Monitoring-Agents/NRPE--
➥2D-Nagios-Remote-Plugin-Executor/details
```

Then untar it and do './configure' followed by 'make all.' For the Nagios server, simply copy check_nrpe to the plug-ins directory and you're done. For the monitored hosts, copy nrpe to somewhere like /usr/sbin and then grab the sample config file from the sample-configs directory and put it in /etc/. Modify the config file to suit your needs; it's heavily commented, so it should be self-explanatory. Finally, 'nrpe –c /etc/nrpe.cfg –d' will launch the NRPE daemon in standalone mode. The sample-configs directory also contains inetd/xinetd configs, if you want to use a superserver instead.

For Microsoft Windows users, your best bet for NRPE-compatible remote execution is the NSClient++ package, available from the following:

```
http://exchange.nagios.org/directory/Addons/Monitoring-Agents/
➥NSClient++/details
```

NSClient++ implements an NRPE listener service and will respond to requests from the check_nrpe plug-in on the Nagios server. More information about NSClient++ is provided in Chapter 6.

End Notes

[1] Some distributions include the plug-ins in their Nagios packages.

[2] If you're going to attempt a Cygwin build, make sure to include –enable-cygwin in your configure options.

[3] www.apache.org.

[4] This has always been the case for me, though my Solaris-adept technical reviewer, Kate Harris, tells me that she had to comment out the SNMP-specific portions of the makefile to get the plug-ins to compile.

[5] Red Hat-specific Nagios RPMs are also available straight from www.nagios.com.

[6] Although manually compiling Nagios on Linux, BSD, and Solaris is straightforward, building it on AIX is not for the faint of heart. If you're going to attempt a local install on AIX, I recommend picking up gcc from bull freeware rather than using AIX's compiler. If you are an AIX 5L user, the affinity program has made building Nagios on AIX much easier.

CHAPTER 4

Configuring Nagios

After installation, Nagios needs some configuration before it can start. Nagios is configured by way of text files that contain directives and definitions. It can seem daunting to configure for the first time because the definitions are self-referential and there's a lot to define.

To get started, Nagios needs to call plug-ins during a time period against hosts and services and send notifications to contacts in the event a check returns a bad status, so you need to define the checks, time periods, hosts, services, notification commands, and contacts (and that's all mandatory). Because so many objects refer to so many other objects, it can be hard to know where to begin to explain it all. Don't be discouraged; Nagios comes with options to generate sample configuration files and, further, you can take a few shortcuts to bootstrap the configuration process,[1] but first you need a good understanding of what a configuration looks like.

Hosts and services are the fundamental objects. They are the objects that refer to most others, so most of the documentation that has been written about Nagios begins with them. I'm going to take more of a bottom-up approach. I'll start by describing the daemon configuration files and then work my way up from commands and time periods, through services and hosts to groups, and then, finally, optional definitions such as escalations and extended information. Each group of objects is referred to by the objects above them, so by explaining things this way, whenever we come to an attribute in a definition that references another object, we will have already looked at what that object consists of. I find that it's easier to get a grasp of the whole picture by explaining it this way.

Objects and Definitions

In Nagios, there are two types of configuration files: those that contain directives and those that contain definitions. Technically, only two configuration files are needed: nagios.cfg and an object configuration file. The nagios.cfg file contains directives that affect the operation of the Nagios daemon: for instance, where and how to write logs, the name of the object config file, global settings, and things of that nature. The object configuration file defines the various objects that Nagios deals with.

There are quite a few different types of objects (as outlined in Table 4.1), so, although it's possible to lump all their definitions into a single file, many people prefer to group the object definitions by type and keep a different file for each type. This makes writing and learning about object configuration easier, so it's the convention I follow throughout this chapter. The nagios.cfg file must be named nagios.cfg. The other configuration file names are user defined, so the filenames I use throughout the chapter are not written in stone. It's entirely possible that you might inherit a Nagios implementation that groups object definitions by network subnet, OS, or even physical proximity to the Pepsi machine. Further, several automated configuration tools like check_mk and NagiosQL use configuration schemes that group objects by host, which is more machine friendly.

Table 4.1　*A Brief Summary of Nagios Objects*

Object Name	Description	Recommended Filename
timeperiod	Timeperiods are the defined blocks of time that other objects use to determine their operational hours and blackout periods.	timeperiods.cfg
command	Command definitions map macros to external programs. Other objects use commands for many things, such as sending notifications and running service checks.	misccommands.cfg and checkcommands.cfg
contact	A contact defines a notification target, which is usually a human being.	contacts.cfg
contactgroup	Contacts are organized into groups called contactgroups. Objects that send notifications always reference contactgroups and never individual contacts. A contact can be a member of any number of groups.	contactgroups.cfg
host	Hosts are physical[2] entities, such as servers, routers, or tape drives.	hosts.cfg

service	Hosts provide one or more services. For a web server, httpd or IIS would be a service. The majority of Nagios configuration is made up of service definitions.	services.cfg
hostgroup	Hosts may belong to any number of user-defined hostgroups. Names and methodology are up to you; for example, servers-with-blue-LEDs or routers-my-boss-refuses-to-upgrade.	hostgroups.cfg
servicegroup	Like hosts, services may belong to any number of user-defined groups. Servicegroups are a feature unique to Nagios 2.0 and later.	servicegroups.cfg
hostdependency	Dependencies filter out checks and notifications for objects, based on the status of other objects. Make sure you read and understand the section "Servicegroups" before using these.	dependencies.cfg
servicedependency	These work the same as hostdependencies.	dependencies.cfg
hostescalation	Escalations provide Nagios the capability to notify additional contacts, such as managers, in the event a problem persists without being acknowledged past a given number of notifications.	escalation.cfg
serviceescalation	These work the same as hostescalations.	escalation.cfg
hostextendedinfo *deprecated*	Extended info objects map titles and graphics to host and service objects for the web interface. These definitions are entirely optional and cosmetic in nature. In Nagios 3.0, extended info attributes have been rolled into the base-level object (hostextinfo -> host). These objects are now deprecated and will not work in future versions later than 4.0.	hostextinfo.cfg
serviceextendedinfo *deprecated*	See the preceding hostextendedinfo.	serviceextinfo.cfg

The nagios.cfg file is required. If you use the web interface, another configuration file, cgi.cfg, is also required. The cgi.cfg file contains configuration directives for the CGIs and is where most of the UI security is configured. The cgi.cfg and the nagios.cfg files contain configuration directives rather than object definitions. The directive syntax should be familiar to anyone who has configured software on a UNIX system. There is one directive per line, followed by an =, followed by the value of the directive. Whitespace is optional and comments begin with a pound (#).

Definitions, no matter the type, use a common syntax that resembles a C function. The definition is composed of a block of directives surrounded by curly braces ({}). The block begins with a define keyword, followed by the object type. Directives within the definition block are whitespace separated, unlike their nagios.cfg counterparts, which use an =. All object definitions have one directive in common: <objecttype>_name.[3] Most also have an alias directive. Comments begin with a pound (#). Listing 4.1 is a host definition example to give you a feel for the syntax.

In Nagios 2.0 and later, it is possible to use regex syntax in place of static text in any directive that accepts a comma-separated list of values.[4] For example, specifying simply * in the host_name directive of a service definition causes that definition to apply to all hosts.

Listing 4.1   *A Sample Host Definition*

```
#A comment about myHost
define host{
    host_name               myHost
    alias                   My Favorite Host
    address                 192.168.1.254
    parents                 myotherhost
    check_command           check-host-alive
    max_check_attempts      5
    contact_groups          admins
    notification_interval   30
    notification_period     24x7
    notification_options    d,u,r
}
```

---

nagios.cfg

Required for daemon start

Refers to: everything

Referred to by: cgi.cfg

If you run make install-config during installation, a nagios.cfg-sample file is written for you. It's specific to the configuration directives you provided, so it should already have the correct locations for lock files, log files, and the like. In fact, for first-time installs, there is usually little you have to change in the nagios.cfg. I recommend you start with an existing nagios.cfg and modify it to suit your needs.

Two things you want to change in the nagios.cfg are the location of your object config files and the check_external_commands directive. There are two ways to specify the location of your object configs. You can either list each object definition file specifically with a `cfg_file` directive, as in Listing 4.2, or you can specify a directory with the `cfg_dir` directive, as in Listing 4.3. If you specify a directory, Nagios parses every file that ends in .cfg in the specified directory.

Listing 4.2   *Specifying Object Config Files Individually*

```
cfg_file=/usr/local/nagios/etc/contactgroups.cfg
cfg_file=/usr/local/nagios/etc/contacts.cfg
cfg_file=/usr/local/nagios/etc/dependencies.cfg
cfg_file=/usr/local/nagios/etc/escalations.cfg
cfg_file=/usr/local/nagios/etc/hostgroups.cfg
cfg_file=/usr/local/nagios/etc/hosts.cfg
cfg_file=/usr/local/nagios/etc/servicegroups.cfg
```

Listing 4.3   *Specifying Object Config Files by Directory*

```
cfg_dir=/usr/local/nagios/etc/
```

Next, if you want the CGI commands in the web interface to work, or you want to use external commands[5] in general, you should tell Nagios to accept external commands from the command file with the following line:

```
check_external_commands=1
```

For external commands to work, the command file must exist in the location specified in the nagios.cfg and its permissions must be set correctly. This is all taken care of by make if you run `make install-commandmode` at install time.

Although those two changes get you up and running, you should be aware of a few directives in nagios.cfg. These break down into two types: global enablers and global timeouts. Table 4.2 describes the global enablers. You should bear these in mind because they enable or disable important features, programwide.

Table 4.2   *Global Enablers in the nagios.cfg*

execute_service_checks	Setting this to 0 turns off service checks programwide. Defaults to 1 (on).
accept_passive_service_checks	Setting this to 0 turns off passive service checks.[6] Defaults to 1 (on).
execute_host_checks	This enables/disables host checks. Defaults to 1 (on).
accept_passive_host_checks	This enables/disables checks of hosts. Defaults to 1 (on).
enable_notifications	This setting controls whether Nagios will send notifications. Defaults to 1 (on).
enable_event_handlers	Event handlers may be globally enabled or disabled. Defaults to 1 (on).
process_performance_data	This determines whether Nagios will check for and handle performance data from plug-ins. Defaults to 0 (off).

Table 4.3 describes the timeouts, which control how long Nagios allows various commands to execute. After a command in the queue is executed, Nagios allows it to run for a user-defined period of seconds. Commands that take longer than that amount of time are killed, so it's important to bear in mind these timeouts if you have a problem with custom checks or event handlers unexpectedly dying. When Nagios kills a command due to timeout, it logs a warning message.

Table 4.3   *Global Timeout Values*

service_check_timeout	The length of time Nagios will wait for a service check plug-in to return its status. Defaults to 60 seconds.
host_check_timeout	The length of time Nagios will wait for a host check plug-in to return its status. Defaults to 60 seconds.
event_handler_timeout	The length of time Nagios will wait for an event handler to finish execution. Defaults to 30 seconds.
notification_timeout	The length of time Nagios will allow a notification command to run. Defaults to 30 seconds
perfdata_timeout	The length of time Nagios will allow a perfdata handler to run. Defaults to 5 seconds.

There's a lot of stuff in nagios.cfg that I didn't mention here, including many of the directives referred to in Chapter 2, "Theory of Operations."

The CGI Config

Not required for daemon start

Refers to: nagios.cfg

Referred to by: nagios.cfg

The cgi.cfg file is the only file, besides nagios.cfg, that contains directives instead of definitions, and unless you are using the web interface, it is optional. The Nagios web interface is very much a separate entity from the Nagios daemon. The daemon has no real knowledge of the existence of the web interface, so it communicates with the daemon via the same mechanisms any other program would: It sends commands to the command file and parses logs and state files for the current state of hosts and services. Therefore, a large part of the directives in the cgi.cfg are there to give the CGI programs that make up the web interface the information they need to send commands to and get information from the Nagios daemon.

Like the nagios.cfg, most of the directives in cgi.cfg shouldn't need to change if you specified them correctly at compile time and built the sample config with make install-config. The directives you want to modify center around the web interface security model, which is pretty simple. The CGIs rely on the web server to handle authentication, so any web server can be used to serve up the web interface, and there is no configuration required for specific users outside of the web server configs.

After a user successfully authenticates, the web interface attempts to correlate the username passed from the web server with a contact in the contacts.cfg. After contacts.cfg is set up, and the web server is configured to authenticate users, the CGIs will allow you to get information about the hosts and services for which you are a contact. This works well for large sites that want least-privilege style security. If you get paged when it goes down, then you are allowed to see it in the web interface, without touching anything whatsoever in the cgi.cfg.

For smaller sites, however, it may be preferable to let everyone see everything. Nearly all sites will want to have a few users, such as the Nagios administrator, who can see everything, whether they are configured as a contact for the host or service in question or not. Table 4.4 shows the directives in the cgi.cfg that make these configurations possible. Most directives take a comma-separated list of users. Each directive that supports a comma-separated list also supports the use of an asterisk (*) to mean all users.

Table 4.4 *Security Related cgi.cfg Directives*

use_authentication	This directive, set to 1 (on) by default, tells the CGIs to use authentication information from the web server. Because the web interface can control how Nagios operates, turning authentication off is a bad idea. If you want everyone to see everything, use default_user_name instead.
default_user_name	Usually set to guest, the default_user_name can be granted permissions that will be inherited by all other users. For example, if jdoe doesn't explicitly have access to see information on a service, but the default user does, jdoe will be able to see the service because all users can see what the default user can. If you want everyone to be able to see all hosts, uncomment this directive and list the default user in authorized_for_all_hosts.
authorized_for_system_information	A comma-separated list of users who are allowed to see information related to the Nagios daemon.
authorized_for_configuration_information	A comma-separated list of users who are allowed to see the contents of the configuration files via the web interface.
authorized_for_system_commands	A comma-separated list of users who are allowed to execute commands relating to the Nagios daemon, such as shutdown and restart.
authorized_for_all_services	A comma-separated list of users who are allowed to see information related to any service that Nagios is monitoring.
authorized_for_all_hosts	Like the above, except for hosts.
authorized_for_all_service_commands	A comma-separated list of users who are allowed to execute commands related to services, such as rescheduling checks and disabling notifications.
authorized_for_all_host_commands	Like the above, except for hosts.

Templates

The brunt of Nagios configuration is made up of object definitions as described in the earlier section "Objects and Definitions." Object definitions involve varying degrees of complexity. Command definitions, for example, are normally composed of no more than two or three lines of text. Service definitions, on the other hand, may contain 31 directives, 11 of which are mandatory. For 100 hosts with 1 service each, that's 1,100 lines of configuration just for service definitions, most of which are redundant. Thankfully, Nagios has a few built-in features that mitigate the need for most of the typing. For example, any of the definitions that refer to hosts may refer to a list of comma-separated hosts instead. Nagios 2 and later allows you to specify a host group instead of a host for some definitions that refer to hosts. Nagios 3 and later allows you to nest groups in other groups. These three features alone bring our 1,100 lines back to 11.

Another wrist-saving feature is template-based configuration. Templates capture redundant directives inside special definitions. Normal objects can then refer to the template and inherit directives instead of specifying them explicitly. Template definitions look and act exactly like their counterparts with two exceptions: the `register` directive and the `name` directive. Listing 4.4 is a templatized version of the host definition from Listing 4.1.

Listing 4.4   *A Host Template and Consumer Definition*

```
# This is my template
define host{
    name                    hostTemplate
    check_command           check-host-alive
    max_check_attempts      5
    contact_groups          admins
    notification_interval   30
    notification_period     24x7
    notification_options    d,u,r
    register                0
}

# myHost is shorter now that it inherits from hostTemplate
define host{
    host_name       myHost
    alias           My Favorite Host
    address         192.168.1.254
    parents         myotherhost
    use             hostTemplate
}
```

As you can see, both definitions define objects of type `host`. The host template, however, has a name directive (instead of a host_name directive) and a `register` directive, which is set to 0. Both `name` and `register` are specific to templates despite the object type, so any other type of template (like a service template) is defined the same way. The `name` directive is self-explanatory; it gives the template a name other objects can refer to. Setting the `register` directive to zero tells Nagios not to treat the object as a host object (don't register it), but rather let other objects inherit settings from it (make it a template).

By adding a `use` directive to myHost's definition, we've instructed Nagios to let myHost inherit settings from the template. The host object will inherit anything that is specified in the template. The host object will override any directives it has in common with the template.

Templates may, in turn, inherit properties from other templates, so it's quite common practice with host definitions to define a global host template, then several OS-specific templates, and then hosts that refer to them. Technically, any object can inherit properties from any other object of the same type, registered or not,[7] to an infinite degree via the use directive. I highly recommend the use of template-based configuration; it's very flexible, saves lots of redundant configuration, and makes for readable config files.

Timeperiods

Required for daemon start

Refers to: none

Referred to by: host, service, contact, hostescalation, serviceescalation

---

Listing 4.5   *Timeperiod Example*

```
define timeperiod{
    timeperiod_name    nonworkhours
    alias              Non-Work Hours
    sunday             00:00-24:00
    monday             00:00-09:00,17:00-24:00
    tuesday            00:00-09:00,17:00-24:00
    wednesday          00:00-09:00,17:00-24:00
    thursday           00:00-09:00,17:00-24:00
    friday             00:00-09:00,17:00-24:00
    saturday           00:00-24:00
}
```

Timeperiods define blocks of time that many other objects reference in the context of operational hours or blackout periods. The timeperiod definition is agnostic; the period of time it defines is not specific to any particular purpose, so two different objects may refer to the same time period for completely different reasons.

Timeperiod definitions have one directive for each day of the week. Omitting a day altogether means the entire day is not included in the time period. Like all other objects, timeperiods may inherit directives from other timeperiods or timeperiod templates. Multiple blocks of time in the same day may be specified by separating them with commas.

In Nagios 3.0, several directives were added to the weekday names to give administrators more flexibility in defining time periods. These are as follows:

Calendar date:

```
2013-01-01        00:00-24:00
```

Specific month date

```
January 1st       00:00-24:00
```

Generic month date

```
Day 15            00:00-24:00
```

Offset weekday of specific month

```
2nd Tuesday in December       00:00-24:00
```

Offset weekday

```
3rd Monday        00:00-24:00
```

Commands

Required for daemon start

Refers to: none

Referred to by: host, service, contact

Listing 4.6 *Command Example*

```
define command{
    command_name    check_ping
    command_line    $USER1$/check_ping -H $HOSTADDRESS$ -w $ARG1$ -c
$ARG2$ -p 5
```

Though composed of only two directives, command definitions are central to the functionality of Nagios. Commands are the means by which Nagios may call external programs and, as we'll see in other definitions throughout the chapter, Nagios calls external programs often.

The most common use of the command object is for calling plug-ins. As previously mentioned, there are only two directives: command_name, which gives the object a name that other objects can reference, and command_line, which defines the shell syntax of the command.

Command objects don't just refer to external programs; they capture the command syntax of external programs. The uppercase words surrounded by dollar signs are called macros. Macros are context-specific internal variables that Nagios replaces at runtime. The HOSTADDRESS macro, for example, refers to the host address of whatever host Nagios happens to be running this plug-in on.[8] This makes it possible to use one command definition for any host. Nagios will fork an exec of command_line exactly as it's written in the definition, but just before executing the command, Nagios replaces all the macros with their actual values.

To avoid unintended shell interpretation and injection attacks, Nagios strips certain characters out of the actual values before it replaces the macro keywords in the commands in some contexts.[9] For the same reasons, Nagios also prevents you from using special characters in host and service names. These characters are user-definable via directives in the nagios.cfg and you should be aware of them so you avoid using them in your definitions. As of the time of this writing, the illegal name characters are

    ` ~ ! $ % ^ & * | ' " < > ? , ( ) =

and the illegal macro output commands are

    ` ~ $ & | ' " < >

Different macros are available in different contexts, so it can be hard to know what macros you can use in every situation. For example, email-related macros, such as CONTACTNAME, are available to commands being run for notifications, but not commands being run for service checks. Check the online documentation at http://nagios.sourceforge.net/docs/3_0/ macros.html for a complete matrix of available macros and the contexts they are available in.

In the nagios.cfg file is a directive called resource_file, which allows you to specify a file to create your own macros. This file is usually called resources.cfg or resource. cfg, and within it you may define up to 32 macros. The resources.cfg usually contains at least one macro, USER1, which resolves to the location of the installed plug-ins. Because resources.cfg is owned by root and read-only, it's a better place to define any usernames or passwords than the checkcommands.cfg, which is world readable. After they are set up in resources.cfg, your command objects can refer to the passwords by macro, thereby keeping them safe from prying eyes.

In Nagios 2.0 and later, macros are exported as environment variables, so any macro available to a command definition is also available to the program called by that definition. In Nagios 3.0 and later, you may also define "custom variables" inside host, service, and contact definitions. Custom variables may be defined by prefacing a directive with an underscore. If, for example, you wanted to track host MAC addresses in your host definitions, you could add the following directive to each host:

```
_macaddr          00:06:5B:A6:AD:AA;
```

To prevent name-space collisions, Nagios will convert your custom variable name to uppercase and prepend "HOST" "SERVICE" or "CONTACT" to it to create a unique macro. For example, the _macaddr custom variable we created below may be referenced as a macro using the name "$_HOSTMACADDR$". When the macro is exported to an environment variable, it will be further transformed into "$NAGIOS_HOSTMACADDR".

Contacts

Required for daemon start

Refers to: command, timeperiod, contactgroup

Referred to by: contactgroup

Listing 4.7 *Contact Example*

```
define contact{
    contact_name                    dave
    alias                           dave josephsen
    host_notification_period        24x7
    service_notification_period     work-hours
    host_notification_options       d,u,r,f
    service_notification_options    w,u,c,r,f
    host_notification_commands      host-email, send-sms
    service_notification_commands   service-email
    email                           dave@somewhere.org
    pager                           555-1024
    address1                        dave_josephsen@gmail.com
    address2                        cn=djosephs,ou=foo,dc=bar,dc=com
}
```

The contact object defines everything Nagios needs to know about a person. The name and alias directives provide the usual. Two timeperiod objects may be specified: the hours during which the contact wants to be notified of host problems and those during which the contact wants to be notified of service problems. Each contact can filter the types of alerts it receives for host and service problems[10] with the `notification_options` directives. The `notification_commands` refer to command definitions that perform notification actions, such as sending emails or controlling armies of semi-autonomous messenger robots. Notification commands usually look something like the one in Listing 4.8. Several addresses can be defined, all of them optional. These include an email address, a pager address, and any number of addressX directives.

Listing 4.8   *A Notification Command Definition*

```
define command{
   command_name    host-notify-by-email
   command_line    /usr/bin/printf "%b" "***** Nagios  *****\n\
nNotification Type: $NOTIFICATIONTYPE$\nHost: $HOSTNAME$\nState:
$HOSTSTATE$\nAddress: $HOSTADDRESS$\nInfo: $HOSTOUTPUT$\n\nDate/Time:
$LONGDATETIME$\n" | /usr/bin/mail -s "Host $HOSTSTATE$ alert for
$HOSTNAME$!" $CONTACTEMAIL$
   }
```

Nagios has no real concept of what these addresses mean; it simply makes them available to the notification commands via macros, as shown in Listing 4.8. It's up to the notification command definition to do something useful with them.

Contactgroup

Required for daemon start

Refers to: contact

Referred to by: host, service, contact, hostescalation, serviceescalation

Listing 4.9   *Contact Example*

```
define contactgroup{
   contactgroup_name    admins
   alias                The Administrators
   members              chris,dave,jason,jer,kelly
}
```

Contacts are organized into groups, and host and service checks refer to the groups rather than individual contacts. The only objects that refer directly to contacts are contactgroups. The group definition is very straightforward, containing the usual `name` and `alias` directives, as well as the `members` directive. In Nagios 2.0 and later, you may specify contact group membership by adding a contactgroups directive to a contact definition instead of maintaining a separate definition for the contactgroup. If you use both a members directive in the contactgroup definition and a contactgroups directive in the contact definition, Nagios will merge the two; however, I recommend you pick one or the other and stick with it. In Nagios 3.0 and later, contactgroup objects have the advantage that they may contain a contactgroup_members directive, which enables you to nest contact groups inside other contact groups.

Hosts

Required for daemon start

Refers to: timeperiod, contactgroup, command

Referred to by: service, hostgroup, hostdependency, hostescalation, hostextinfo

---

Listing 4.10   *Host Example*

```
define host{
    host_name               myHost
    alias                   My Favorite Host
    address                 192.168.1.254
    parents                 myotherhost
    event_handler           ups-reboot
    check_command           check-host-alive
    max_check_attempts      3
    contact_groups          admins
    notification_interval   30
    notification_period     24x7
    notification_options    d,u,r
}
```

Because hosts and services are really the central objects in Nagios, their definitions are more involved than most. You may specify 28 different directives within a host definition, but only nine of them are mandatory.[11] Additionally, in Nagios 3.0 and later, you can make up your own directives—called "custom variables"—as I described earlier in the "Commands" section. See the Nagios documentation at http://nagios.sourceforge.net/docs/3_0/xodtemplate.html#host for a complete list of the available directives. The directives in Listing 4.9 are probably all you'll need.

The address directive tells Nagios how to find the host. You may specify either an IP address or a hostname for the address directive, and this is an important decision. If you specify hostnames, DNS problems will cause host failures because Nagios will be unable to resolve the hostname. On the other hand, if you specify an IP address, you have to remember to change the definition in the event that the IP of the host changes. In large environments, this may happen pretty often and without your knowledge. I prefer to specify hostnames and to run a local DNS name server service, such as tinydns,[12] locally on the Nagios box. Using a local name cache solves most DNS-related issues because Nagios uses itself as a name server, but it also necessitates some type of replication with the real name servers.

The `parents` directive tells Nagios where in the network topology that the host resides. Parents are defined from the perspective of the Nagios server. If the Nagios server is connected to a router, which also connects to a separate subnet containing four hosts, each of those hosts should be defined with a `parents` directive listing the router. The router does not need a `parents` directive, because it is on the same subnet as the Nagios server. Each host may list one or more parents. As described in Chapter 2, Nagios uses parent/child relationships to treat outages on remote subnets differently from those on local subnets. If a host with children, such as a router or switch, goes down, Nagios will consider the router down and its children unavailable. This is an important distinction for reporting and notifications. As shown in the "Contacts" section, the contact may filter out the unavailable notifications via its notification options directives.

The `event_handler` and `check_command` directives both specify command objects. Event handlers, as described in Chapter 2, are usually commands that launch scripts in an attempt to rectify simple problems automatically. When a host changes state, Nagios executes the host's `event_handler` before sending notifications.[13] The `check_command` is the command Nagios uses to check that the host is available. The command definition usually points to the check_ping plug-in. As described in Chapter 2, Nagios only runs the `check_command` if a service check fails on the host, so while directives exist to enable regularly scheduled checks of the host, these are discouraged. By design, Nagios will run `check_command` automatically as needed.

If Nagios runs the `check_command` and it fails, it will place the host in a soft down state and retry the `check_command` as many times as specified by `max_check_attempts`. If the `check_command` fails each time, the host is placed in a hard down state, notifications are performed, services on the host are assumed down, and service-related checks and notifications on the host are postponed until the host `check_command` returns an okay state. Setting `max_check_attempts` to 1 effectively disables soft states for the host. If the `check_command` fails once, it is immediately placed in a hard down state and notifications are sent.

The last four directives are notification-related and answer the questions "Who?," "What?," "When?," and "How often?" In the event that the host changes into one of the hard states specified by the `notification_options` directive, and that change occurs during the time period specified by the `notification_period`, notifications will be sent to the groups specified by the `contact_groups` directive. For problem states, follow-up notifications will be sent every so often until the host recovers. Exactly how often is specified by the `notification_interval`.[14]

Services

Required for daemon start

Refers to: host, timeperiod, contactgroup, command

Referred to by: servicegroup, servicedependency, serviceescalation, serviceextendedinfo

---

Listing 4.11   *Service Example*

```
define service{
    host_name                  myServer
    service_description        check-disk-sda1
    check_command              check-disk!/dev/sda1
    max_check_attempts         5
    normal_check_interval      5
    retry_check_interval       3
    check_period               24x7
    notification_interval      30
    notification_period        24x7
    notification_options       w,c,r
    contact_groups             admins
    }
```

Service objects glue it all together. They refer to every mandatory object, defining the specifics of how you want to run a given plug-in, where you want to run it, how often, and whom to call when things go wrong. A whopping 31 directives may be specified in a service definition. The 11 shown in Listing 4.11 are mandatory. The remaining 20 can be found in the online documentation at http://nagios.sourceforge.net/docs/nagioscore/3/en/objectdefinitions.html#service.

The `host_name` directive specifies a comma-separated list of hosts on which this service runs. The service definition breaks with the name/alias convention in favor of a single `service_description` directive. This is because, unlike the other objects, services aren't required to have unique names; they only need to specify a unique set of hosts. So it's perfectly fine to create multiple service objects with the same name but completely different settings, as long as they don't share a reference to the same host object. This is handy, for example, when more than one host needs the same service with a different retry interval. In this situation, the service can be copied, or referred to with a use directive with a different host_name and interval specified.

Like host objects, the check_command directive specifies the command object used to check the service. It's common for plug-ins to provide a subset of functionality;[15] for example, instead of having two plug-ins called check_sda1, and check_sda2, the plug-ins tarball has a single plug-in called check_disk, which is capable of checking any disk. The check_disk plug-in takes the name of the disk as an argument on the command line and checks it.

Service objects, on the other hand, tend to be single purpose, and our service in Listing 4.10 is no exception. It uses the check_disk command to check a single disk, namely /dev/ sda1. Because command syntax may contain whitespace, an exclamation mark is used to separate the command name from the arguments we want to pass to it. Each argument is made available to the command object via a numbered 'ARG' macro. In our preceding example, when Nagios dereferences the check_disk command, it replaces the command definition's $ARG1$ macro with /dev/sda1, and exec the resulting command. Any number of exclamation mark-separated arguments is supported.

Service notifications are a bit more straightforward than host notifications, but they follow the same basic pattern. Service checks that return bad statuses are retried a number of times to ensure they are down and remain down. While Nagios is verifying the state of a service with retries, the service is placed in a soft state. When the service is verified to be down, it is placed in a hard state and notifications are sent. Follow-up notifications are sent every so often until the service recovers.

All _interval type directives in Nagios refer to a number of time units to wait before doing something. What a time unit means is user definable via the interval_length directive in the nagios.cfg. By default, this directive is set to 60 seconds, so, in general, any interval definition is going to refer to the number of minutes to wait before doing something.

max_check_attempts is the number of times Nagios will retry a service. normal_ check_interval is the number of minutes to wait between service checks. retry_check_ interval is the number of minutes to wait between checks when the service is in a soft state and Nagios is trying to verify the service state. The time period within which checks may be scheduled is given by the check_period. The notification_interval specifies the number of minutes to wait between follow-up notifications. notification_period defines the time period within which notifications may be sent. notification_options filters the type of notifications this service will send[16] and, finally, contact_groups specifies to whom the notifications should go.

Based on the preceding information, it should be obvious that the amount of time a service will spend in a soft state is a function of max_check_attempts and retry_check_ interval. If you aren't getting notifications quickly enough, you can either retry less or lessen the amount of time between retries.

Hostgroups

Required for daemon start

Refers to: host

Referred to by: host, hostescalation

Listing 4.12 *Hostgroup Example*

```
define hostgroup{
    hostgroup_name        oracle-servers
    alias                 Servers Running Oracle
    members               server1,server2
    }
```

Identical in syntax to contactgroups, hostgroup objects exist to ease administration and reporting. Hosts may belong to multiple host groups. Membership may be defined in the host objects via the `hostgroup` directive instead of using members in this object. A unique feature of host groups is that multiple `members` directives may be defined.

In Nagios 3.0 and later, `hostgroup` definitions may contain a `hostgroup_members` directive, enabling you to nest hostgroups inside other hostgroups.

Servicegroups

Not required for daemon start

Refers to: service

Referred to by: service

Listing 4.13 *Servicegroup Example*

```
define servicegroup{
    servicegroup_name     disks
    alias                 Disks
    members               myServer,chk-disk,server1,chk-disk
    }
```

Optional and similar to `hostgroup` definitions, service groups are new to Nagios 2.0 and are mostly used by the CGIs of the web interface. The syntax of the `members` directive is different from the other group types, listing first a host object followed by the corresponding service object, separated by commas. In Nagios 3.0 and later, `servicegroup` definitions may contain a `servicegroup_members` directive, enabling you to nest service groups inside other service groups.

Escalations

Not required for daemon start

Refers to: host, service, hostgroup, contactgroup, timeperiod

Referred to by: <nothing>

Listing 4.14 *Servicescalation Example*

```
define serviceescalation{
    host_name               myServer
    service_description     check-disk-sda1
    first_notification      4
    last_notification       0
    notification_interval   30
    contact_groups          admins,themanagers
    }
```

Listing 4.15 *Hostescalation Example*

```
define hostescalation{
    host_name               router-34
    first_notification      5
    last_notification       0
    notification_interval   60
    contact_groups          routeradmins,admins
    }
```

It's possible to configure Nagios to involve managers or other techs, in the event that a problem persists beyond a certain number of notifications without being acknowledged. Nagios does this by way of escalation definitions. Escalations can be configured for hosts or services, and the definition syntax is nearly identical for each. The two main differences are that host escalations may specify host groups instead of hosts, and service escalations must specify the host and the service.

Each time Nagios decides to send a notification, it first checks to see if any escalation definitions match the notification it is about to send. If an escalation definition matches the notification Nagios wants to send, Nagios sends the escalation instead. It's important to note that this is an either/or proposition, meaning that if the service definition specifies that the admins contact group be notified and the escalation specifies that the managers be notified, the escalation will win and the admins contact group will get nothing, so be sure to include everyone that needs to be notified in the escalation definition.

The `first_notification` directive specifies the notification number for which the escalation is first enabled. For example, Listing 4.14 matches the fifth notification of a host down alert for router34. The first four notifications will be sent as normal, but the fifth will be an escalation. Escalation notifications will continue to be sent until the number specified by `last_notification`. If the `last_notification` directive is set to 0, escalations will continue until the host becomes available again.

It is possible to have multiple escalations that match the same notification. If this happens, Nagios sends both escalations, so if two escalations match the same notification and have different `contact_groups` directives, all the contacts are notified.

Escalations may specify a custom `notification_interval`, which defines the amount of time to wait between notifications. This interval takes precedence over the interval originally specified in the service definition. In the event that two escalations match one notification, and the escalations contain different `notification_interval` settings, Nagios will pick the smallest interval and use it.

Dependencies

Not required for daemon start

Refers to: host, service

Referred to by: <nothing>

Listing 4.16 *Hostdependency Example*

```
define hostdependency{
    host_name                    myHost
    dependent_host_name          server1
    notification_failure_criteria d,u
    }
```

Listing 4.17   *Servicedependency Example*

```
define servicedependency{
    host_name                        NAS1
    service_description              PING
    dependent_host_name              myServer
    dependent_service_description    check_httpd
    execution_failure_criteria       w,u,c
    notification_failure_criteria    w,u,c
    }
```

Dependencies exist to capture services and hosts that rely on each other. If, for example, you have some web servers using a network-attached shared storage back-end, you can make the web server services dependent on the NAS server's ping service. In the event that the NAS box becomes unavailable, Nagios handles the notifications accordingly and reports will reflect the outage in the context of the web servers. Although host and service dependencies may be defined, it is almost always preferable to use parent/child relationships to capture interdependencies between hosts. Only host definitions may contain `parents` directives, so defining interdependent services requires a service dependency object.

Host dependencies require only `host_name`, which specifies the host that is depended upon, and `dependent_host_name`, which is self-explanatory. Service dependencies obviously must specify the hosts and services in question. The services are specified by way of the `service_name` and `dependent_service_name` directives.

Before Nagios checks the state of a service, it first checks the state of all the services that the service depends upon (its parents). If all of those services are okay, Nagios proceeds to check the child service. If any of the parent services are down, Nagios assumes the child service is down as well and stops checks and notifications on the child. It's possible to modify this behavior by way of the `execution_failure_criteria` and `notification_failure_criteria` directives. These directives can be confusing because they specify when something should *not* happen.

The `execution_failure_criteria` directive specifies the situations in which the child service should *not* be checked. For example, specifying a "w" for warning here means Nagios should not schedule checks of the child service when the parent is in a warning state, which is probably what you want. Setting this to "n" for none would mean that active checks always take place, no matter the state of the parent service, and setting an "o" for okay here would mean that active checks would not take place, even if the parent is in an okay state.

Similarly, the `notification_failure_criteria` directive specifies the situations for which notifications should *not* be sent out for the child service. Like execution_failure_criteria, the options specify states of the parent host. Setting c, for example, means that notifications for the child should not be sent if the parent is in a critical state. The same options are available to both directives: okay, warning, critical, unknown, pending, and none, for services; okay, down, unreachable, pending, and none, for hosts.

Extended Information

Not required for daemon start

Refers to: host, service

Referred to by: <nothing>

Listing 4.18 *Hostextendedinfo Example*

```
define hostextinfo{
    host_name      myServer
    notes          this is my server.. many like it.. yadda yadda
    notes_url      http://foo.com/hostinfo.pl?host=myServer
    icon_image     linux40.png
}
```

Extended information can be defined for hosts and services. This is optional information that is used by the CGIs of the web interface to do things, such as draw pretty icons. All directives are optional, except for `host_name`, which can be a comma-separated list of hosts. The notes_url provides an easy way to link from the web interface to external sites for host information. The icon_image directive specifies a 40x40 pixel image to use to represent this host whenever it appears in the web interface. Icon sets for this purpose can be downloaded from the Nagios Exchange.[17] The icons are expected to be in the webroot/images/icons directory.

Apache Configuration

After Nagios has been configured, the web server must be configured to serve up the web interface's content. The majority of Nagios installations use the Apache web server to serve up the interface, so the configs in Listing 4.19, shamelessly stolen from the official Nagios documentation, are for the Apache web server.

Listing 4.19   *Apache Sample VirtualHost Config*

```
ScriptAlias /nagios/cgi-bin /usr/local/nagios/sbin

<Directory "/usr/local/nagios/sbin">
    Options ExecCGI
    AllowOverride None
    Order allow,deny
    Allow from all
    AuthName "Nagios Access"
    AuthType Basic
    AuthUserFile /usr/local/nagios/etc/htpasswd.users
    Require valid-user
</Directory>

Alias /nagios /usr/local/nagios/share

<Directory "/usr/local/nagios/share">
    Options None
    AllowOverride None
    Order allow,deny
    Allow from all
    AuthName "Nagios Access"
    AuthType Basic
    AuthUserFile /usr/local/nagios/etc/htpasswd.users
    Require valid-user
</Directory>
```

The directives beginning in Auth provide simple login functionality via a text file called htpasswd.users, which resides in /usr/local/nagios/etc/. The htpasswd program, provided with Apache, can be used to create the htpasswd.users file like so:

```
htpasswd -c /usr/local¹⁸/nagios/etc/htpasswd.users dave
```

After the file is created, more users can be added in the same manner, but make sure to drop the –c switch after the first user, because it is used to create the file and will overwrite any files that already exist. As described in the "The CGI Config" section, users who authenticate through the web server via the htpasswd.users file are matched with contacts in the contacts. cfg that have the same name.

It's common for people to get the web interface working but then to have trouble with the CGI commands. For the CGI commands to work, the user ID used by Apache must be a member of the group used for Nagios commands. This group is specified to the configure script at compile time and defaults to nagios. An excellent primer on configuring Apache with Nagios can be found at http://nagios.sourceforge.net/docs/3_0/cgiauth.html.

GO!

At this point, Nagios is ready to start. You can call Nagios with a -v switch to check the config files for errors, like so:

```
/usr/local/nagios/bin/nagios -v /usr/local/nagios/etc/nagios.cfg
```

The Nagios daemon will start in interactive mode, check the config files for errors, and provide a helpful summary screen. If no errors are present, Nagios may be started via its init script. Congratulations! That wasn't such a chore, now, was it?

End Notes

[1] See Chapter 5, "Bootstrapping the Nagios Config Files."

[2] Or the virtual representation of physical entities for those of you who use virtualization such as Xen or VMware.

[3] For example, a host object has a host_name directive, whereas a servicegroup has a servicegroup_name directive.

[4] This option is disabled by default and must be enabled by setting the use_regexp_ matching directive to 1 in the nagios.cfg.

[5] As described at the end of Chapter 2.

[6] Check Chapter 2 for a detailed discussion of passive checks.

[7] I don't recommend having normal objects inherit from other normal objects. Stick to dedicated templates or things can get muddled quickly. Consider yourself warned.

[8] See the "Services" section for a description of the ARG macros.

[9] Host and service notifications and escalations, but not host and service checks.

[10] As discussed in Chapter 2: down, unreachable, recovered, and flapping for hosts.

[11] All of the directives in Listing 4.10 are mandatory, except for event_handler and parents.

[12] www.tinydns.org

[13] Nagios does not wait for the event handler to return before it sends notifications. However, because event handlers are executed at soft state changes, they usually have a window of time to do their work before notifications are sent, while Nagios retries the checks. See Chapter 2 for a discussion of hard versus soft states.

14 Technically, the notification_interval specifies the number of time units to wait between notifications. The interval_length directive in the nagios.cfg specifies the number of seconds in a time unit. The default is one minute, so a notification_interval of 30 equates to 30 minutes unless you've changed it.

15 Therefore, it's also common for command objects to provide a subset of functionality.

16 As discussed in Chapter 2, the possible states are warning, unknown, critical, recovered, and flapping for services.

17 www.nagiosexchange.org

18 The specific init syntax for your system may vary, but something similar to /etc/init.d/nagios start should do the trick.

## CHAPTER 5

Bootstrapping the Nagios Config Files

Everyone seems to have a different idea of what makes for ease of use when it comes to configuration, but most, if the blogosphere is any indication, seem to agree that Nagios isn't it. Although I've yet to meet anyone who enjoys configuring Nagios from scratch, it's sometimes surprising to me what people think will rectify the situation. Web forms, databases, autodiscovery tools, and wizards have all been proposed and created, resulting in a dizzying array of options that seek to simplify the act of configuration. So many, in fact, that I would be remiss if I didn't include a chapter on making configuration easier.

Depending on the size of your environment, there is a point of diminishing returns with tools like these; sometimes learning the tool is more trouble than editing the Nagios config files manually. Where that point resides you'll have to judge for yourself, but in this chapter, I cover three increasingly complex methodologies to help you bootstrap the configuration process. I start with simple scripting templates, move on to automatic discovery and configuration tools, and finish with NagiosQL, the most popular database-backed Web front-end.

The tools all have their good points and bad. For example, some might think the shell scripts lack user friendliness, but on the other hand, PHP raises the vulnerability footprint of the Nagios server. None of these tools are mutually exclusive; it's entirely possible to mix and match them to get what you want.

Scripting Templates

Many Nagios administrators (myself included) maintain a set of "bootstrap templates." These aren't object templates of the type discussed in Chapter 2, "Theory of Operations," and Chapter 4, "Configuring Nagios." They are skeleton config files that can be easily combined with lists of hosts to create valid Nagios configurations. To avoid confusion with object templates, I refer to them as skeletons for the remainder of the chapter.

With this methodology, all that's needed in practice is a plain-text list of hosts. After the skeletons are created, we loop through the host list, using sed (or whatever) to inject the hostnames into the skeletons, thereby creating valid definitions.

There are usually two skeletons for each type of object: an object template skeleton and an object definition skeleton. The template skeleton is a Nagios object template that encompasses as much general information as possible. The definition skeleton has just enough information to define an object and relies on the template for everything else. We can then specify more information, on a per object basis, to override the overly general template as necessary. Listing 5.1 is an example of a host template for use with a definition skeleton.

Listing 5.1   *A Host Template Skeleton*

```
define host{
name                        generic_host
max_check_attempts          2
notification_interval       60
check_period                24x7
notification_options        d,u,r
check_command               check-host-alive
contact_groups              admin
register                    0
}
```

As you can see, the host template skeleton is a normal object template in Nagios. The only arguably odd thing about it is that contact_groups is a directive people usually assign on a per-host basis. Take this template and put it in a file called 'hosts.cfg'. Listing 5.2 shows the second half of the equation, the definition skeleton.

Listing 5.2   *The Host Definition Skeleton*

```
define host{
use            generic_host
host_name      NAME
alias          NAME
address        NAME.DOMAIN
}
```

Given those listings, you probably have a pretty good idea of how this is going to shake out. The skeleton host object will inherit everything it needs from the generic_host template, except for the name and address of the host. Save the object skeleton in a file such as "hosts. skel". All that's missing is a hostname and a domain name. We can easily fill this in with a list of hosts, such as the one in Listing 5.3.

Listing 5.3   *A List of Hosts*

```
Host1.mydomain.com
Host2.mydomain.com
Host3.mydomain.com
```

Finally, the shell script in Listing 5.4 ties it all together. It's little more than a one-liner while loop that takes a list of hosts from standard input. Lines 5 and 6 extract the host and domain names from each element of the list and line 7 replaces the keywords 'NAME' and 'DOMAIN' in the skeleton file with their actual equivalents.

Listing 5.4   *A Shell Script to Create a hosts.cfg from the Skeletons and Host List*

```
#!/bin/sh

while read i
do
    NAME='echo ${i} | cut -d. -f1'
    DOMAIN='echo ${i} | cut -d. -f2-'
    sed -e "s/NAME/$NAME/" -e "s/DOMAIN/$DOMAIN/" hosts.skel >>
➥hosts.cfg
done
```

You can do the same thing with the services.cfg and hostgroups.cfg. Services.cfg will probably specify little other than a ping service. The template would look something like the one in Listing 5.5.

Listing 5.5   *A Services Template for Use with a Definition Skeleton*

```
define service{
    name                  generic-service
    active_checks_enabled         1
    passive_checks_enabled        1
    parallelize_check             1
    obsess_over_service           0
    check_freshness               0
    notifications_enabled         1
    event_handler_enabled         1
    flap_detection_enabled        0
    process_perf_data             1
```

```
        retain_status_information        1
        retain_nonstatus_information     1
        check_period                     24x7
        max_check_attempts               2
        normal_check_interval            5
        retry_check_interval             1
        notification_interval            60
        notification_period              24x7
        notification_options             w,u,c,r
        contact_groups                   admin
        register                         0
}
```

With so much defined in the template, the same shell script in Listing 5.4 can be used with the definition skeleton in Listing 5.6 to create valid ping services for each host.

Listing 5.6   *A Services Definition Skeleton*

```
define service{
        use                      generic-service
        host_name                NAME
        service_description      PING
        notification_options     c,r
        check_command            check_ping!500.0,20%!1000.0,60%
        }
```

Hosts and service definitions are the hard part. After you are done with those, you can define the "admin" group in contactgroups.cfg, define the members of the group in contacts.cfg, create a hostgroup with all the hosts, and you're ready to go. Of course, each of those files can be scripted with skeletons in much the same way for larger installs.

I once worked with someone who told me that I was the type of person who liked to "cut metal." What he meant by that was that I wanted to begin implementing before the planning was done . Although I'm familiar with several ways to bootstrap the Nagios configs, I always seem to come back to using skeletons, because they allow me to get some work done and adhere to the "Procedural Approach to Systems Monitoring" I describe in Chapter 1, "Best Practices," at the same time.

In fact, I use Nagios templates and skeletons to document business requirements during the planning process. By the time implementation time rolls around, I usually have some complex configuration out of the way and just need a list of hosts and a few shell scripts. I find this makes implementation a breeze, gets everyone what they expect from the beginning, and most important, allows me to at least "put chalk lines on the metal." So if you have itchy tinsnip fingers, and some shell chops, you can't go wrong with skeletons.

Autodiscovery

When I wrote the first edition of this book in 2007, the capability of a monitoring system to perform "autodiscovery" (the automatic configuration of hosts and services based on output from a network scanner) for whatever reason, seemed to be *the* feature that enabled forum trolls to distinguish the "cool" monitoring systems from the insipid wanna-be toys. Nagios, being bereft of a built-in autodiscovery capability (as well as built-in everything else for that matter) put it in the latter group. It seemed at the time that I couldn't read a monitoring-related forum without involuntarily rolling my eyes at the marketeers for various commercial products who were forever chanting this strange "autodiscovery or death" rhetoric. Why they chose that particular feature I can't guess. I've rarely in my professional career found myself in want of such a tool for Nagios. I usually find it more expedient to hack together whatever suits the situation by combining NMAP along with the skeleton templates I described in the previous section.

That's not to say that Nagios lacked options for autodiscovery. On the contrary, the whining in the forums beget an explosion of these tools in every language at every level of complexity, so when I wrote this section in the first edition of this book, my challenge was to narrow the field to a few tools I thought might be useful to real administrators trying to do real work. This was a real challenge because, as a group, I find these tools to be unwieldy; they tend to make strange assumptions and are overly enamored of XML. Anyway, it just never seemed to me to be the kind of thing one went looking for tools to solve; it seems to me like the kind of thing a systems administrator just *does*.

Fast forwarding to 2012, as I undertake to ensure that this chapter is still relevant, I find that the autodiscovery tools have either disappeared or have become abandonware. Yes, *all* of them. 100%. Further, some light googling retrieves only ancient blog posts from bygone tool-writers announcing or justifying the creation of their now-abandoned hot new autodiscovery tool for Nagios (now with extra XML!). Given the firestorm of controversy that once surrounded this topic, I find it disorienting that not only the tools but even the trolls have utterly vanished. This is a vexing turn of events for the "Autodiscovery" section of this chapter, but not, I think, for the Nagios community, I suspect two things account for it.

Check_MK

The first is a plug-in written by Mathias Kettner called Check_MK, which I cover at length in Chapter 6, "Watching: Monitoring Through the Nagios Plug-in," and again in Chapter 7, "Scaling Nagios." Check_MK is not an autodiscovery tool as such, but rather a self-contained client-side monitoring agent that happens to work very well with Nagios. If you're looking for an easy way to bootstrap a new Nagios Core system, it doesn't get much easier than Check_MK. After the MK Agent is installed on the hosts in your network, they are not only automatically detected and configured, but a whole slew of services—everything you want to know, and probably more—on your hosts is detected and set up. MK does not use autodiscovery via network scanning, but rather service-discovery via interprocess communication, so it is able to automatically configure Nagios to check services like CPU and RAM utilization. It even automatically graphs all the performance data from the services it sets up. All you need to do is install the MK Agent and run the MK discovery tool on the server. Check_mk handles the rest.

Nagios XI

The second is the commercial version of Nagios, which was released in 2009. Nagios XI is everything the trolls would expect to see in a "cool kid" monitoring system and more. This includes not only prepackaged autodiscovery (via network scanning) features, which are slick looking and effective, but also configuration wizards, which are designed to insulate end users from the intricacies of the Nagios Core configuration syntax (another pet subject of monitoring forum trolls).

Autodiscovery Is Dead: Long Live Autodiscovery

I can't prove it, but my suspicion is that real administrators with real problems to solve discovered check_mk and never looked back, or convinced their managers to pony up and buy them XI (or both). At the same time, one look at the XI screenshots caused a massive spontaneous troll migration away from the monitoring forums and toward dpreview.com or perhaps YouTube, where they all live happily trolling it up to this day (sorry about that, YouTube).

I jest—but truly, I think my hypothesis has some merit. It certainly accounts for the loss of interest in standalone autodiscovery tools for Nagios. There simply isn't room for them anymore. If you're the kind of sysadmin who likes to get hacky with Nagios Core, you're going to write a one-liner for autodiscovery and be done (the old autodiscovery tools wouldn't have given you enough control, anyway). If you're the type who just wants to install something without getting too involved, you'll install check_mk and be done. And if you're in the market for an effective, established, polished commercial product with support behind it, you'll buy Nagios XI and be done.

NagiosQL

NagiosQL is a Web-based configuration tool for Nagios. As depicted in Figure 5.1, it is implemented as a PHP program that runs from a Web server with PHP support like Apache. The front-end PHP program takes configuration changes from the user and writes them to a MySQL database. Back-end scripts then export the configuration from MySQL and generate configuration files that are read by the Nagios daemon.

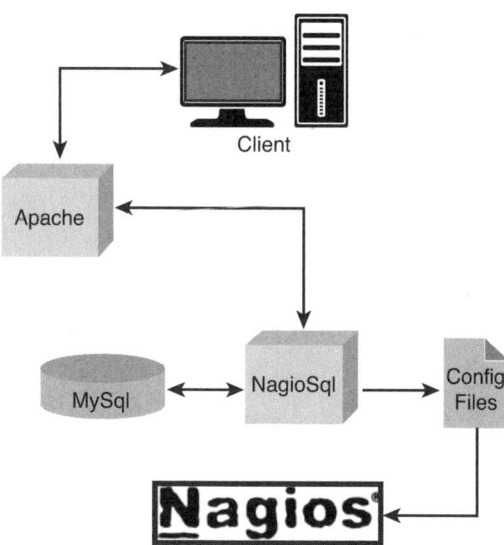

Figure 5.1 NagiosQL architecture

NagiosQL is a well written and actively developed production-grade tool. It is, in fact, the engine that powers the "Advanced" configuration features in Nagios XI, the commercial version of Nagios. I like the approach NagiosQL has taken because it doesn't really modify the Nagios daemon in any way. Nagios still reads its configuration from files on the hard drive, so you don't need to do anything special to back up the configuration or restore it if something goes wrong. NagiosQL includes tools that you can use to import your existing configuration files; it also allows you to maintain your own static configuration separate from its database-driven config.

The capability to maintain your own static config in a way that's separate from the Web-based configuration tool is important because it's dangerous to assume that all configuration will be created using the Web interface. The check_mk plug-in I mentioned in the previous section is a great example of a tool you might want to use in addition to Web-based configuration to make your life easier. Many of the Web interfaces for Nagios I've dealt with over the years would lock you out of this type of integration by modifying Nagios to read configuration straight from the database.

NagiosQL can generate any kind of Nagios configuration, including templates, and can even modify non-object configuration like the nagios.cfg and cgi.cfg. Compared to configuration front-ends I've used in the past, I find its interface (depicted in Figure 5.2) clean and intuitive. Indeed, it presents the Nagios configuration options in a manner that makes it seem like there isn't really much involved in configuring Nagios at all.

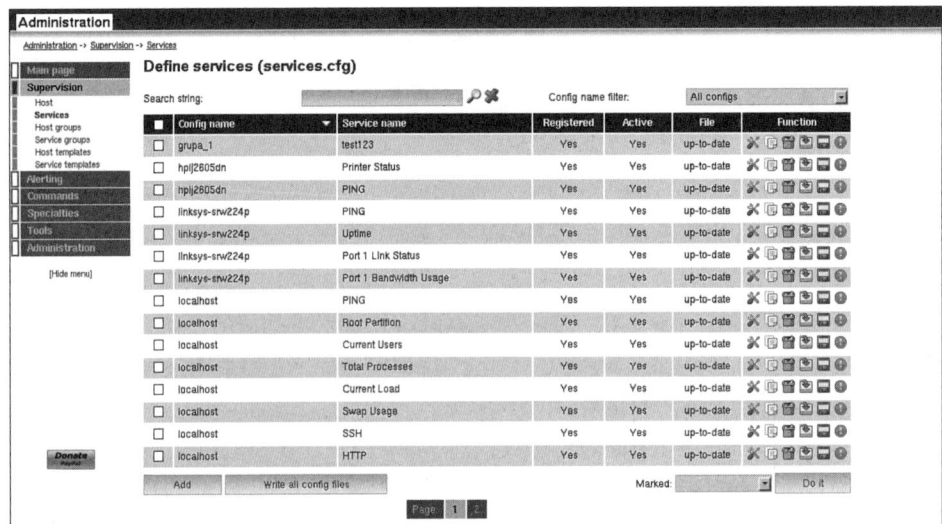

Figure 5.2   The NagiosQL services configuration screen

The install is straightforward—simply download and extract into your Web directory. You'll need a working Apache/PHP/MySQL environment on your Nagios server, and you'll have to modify your nagios.cfg file to point at NagiosQL's configuration files. The full installation instructions may be found here:

www.nagiosql.org/documentation.html#NagiosQLx

Watching: Monitoring Through the Nagios Plug-ins

Welcome to where the rubber meets the road. Until now, I've talked a lot about how Nagios works and how it is installed and configured. This chapter ties together the theoretical work of the previous chapters and talks about the nitty gritty of performing systems monitoring. As I've repeatedly said throughout the book, Nagios is a scheduling and notification framework. Small, single-purpose programs called plug-ins do the monitoring. So an in-depth discussion about monitoring will mostly be a discussion about plug-ins.

Nagios is limited only by the availability of plug-ins for a given task. Writing your own plug-ins, as described in Chapter 2, " Theory of Operations," is trivial and highly encouraged, so thousands of them are available. This chapter should give you a good basis to understand what types of plug-ins are available, how to use them, and where to begin when you need to start making your own. The chapter is broken into four parts, mostly along architectural lines, with a section each for local queries, Microsoft Windows, UNIX, and "Other Stuff," which includes networking gear and environmental sensors.

Local Queries

There are three primary ways to monitor hosts and services with Nagios. Nagios can use various remote execution techniques to connect to remote hosts and to run plug-ins located therein. Nagios can wait in a passive mode for remote hosts to notify it of trouble by defining passive checks. Finally, Nagios can launch plug-ins locally to query the availability of various hosts and services from afar. Let's start the chapter by checking out some of the things Nagios can do without leaving home.

Pings

By far, the most common service check in any monitoring program is the classic ICMP echo request, more commonly known as the ping. Although pings are easy to set up in Nagios, they can be somewhat confusing to first-time users of Nagios because, at first glance, they appear to be overused or redundant.

At a minimum, Nagios requires that there be at least one service per host. The most common service to set up first is a ping service because it is a very simple definition, and if you're using the example configs, there is already a ping check defined for you. If you read Chapter 4, "Configuring Nagios," you know that each host definition also contains a check_command directive that specifies the command to use to verify that the host is operational. This usually defaults to the check_host_alive command, which, in turn, uses ping. So why have two ping checks per host? It's important to remember that, as described in the "Scheduling" section of Chapter 2, host check commands are not scheduled unless they are needed. In practice, the host check command will not be run until a service check has already failed. Thus, although the definitions may seem redundant, the checks exist for very different purposes.

Let's walk through setting up a simple ping check. Assuming the hosts and contacts are already defined, the first step is to define the command in the check_commands file. Command definitions, as described in Chapter 4, "Configuring Nagios," glue service definitions to external monitoring programs. The purpose of the check command is to tell Nagios what external command to launch and how. A command definition for ping might look like Listing 6.1.

Listing 6.1   *Check_ping Command Definition*

```
define command{
    command_name    check_ping
    command_line    $USER1$/check_ping -H $HOSTADDRESS$ -w $ARG1$ -c
$ARG2$ -p 5
    }
```

Listing 6.1 is a classic example of a Nagios command definition. They do get simpler than this, but not by much. The –H switch is nearly always used to specify the hostname or IP of the box against which to run the plug-in. Likewise, -w and -c nearly always specify the warning and critical thresholds. The check_ping plug-in uses the -p switch to specify how many ping packets to send.

The words surrounded by dollar signs are macros. Macros are described fully in Chapter 4, but suffice it to say that when Nagios calls check_ping, it will first replace the macro names with real values from various places. For example, the numbered ARG macros will be replaced with values from arguments in the service definition. Speaking of which, a service definition such as the one in Listing 6.2 is the next step in setting up our ping check.

Listing 6.2   *Check_ping Service Definition*

```
define service{
    host_name               Server
    service_description     check_ping
    check_command           check_ping!500.0,20%!1000.0,60%
    max_check_attempts      5
    normal_check_interval   5
    retry_check_interval    3
    check_period            24x7
    notification_interval   30
    notification_period     24x7
    notification_options    w,c,r
    contact_groups          admins
    }
```

Again, I'll refer you to Chapter 4 for the specifics and summarize here. This service defines that the check_ping command will be run against Server every 5 minutes, for 24 hours a day, 7 days a week. If something goes wrong, Nagios will verify that the problem persists by rechecking the service four more times, 3 minutes apart. Once verified, Nagios will begin notifying the admins group every 30 minutes, for 24 hours a day, 7 days a week, until the service comes back online.

Many of the options in Listing 6.2 would be better specified inside a template, so in the interest of brevity, the rest of the service definitions in this chapter assume that you use a template such as the one in Listing 6.3. Check out the "Templates" section in Chapter 4 for a discussion of templates and how to use them.

Listing 6.3   *The Template Used by the Rest of the Examples in This Chapter*

```
define service{
    name                    chapter6template
    max_check_attempts      5
    normal_check_interval   5
    retry_check_interval    3
    check_period            24x7
    notification_interval   30
    notification_period     24x7
    contact_groups          admins
    register                0
}
```

The funky-looking stuff after check_ping in Listing 6.2's check_command directive is a list of arguments. Arguments in service definitions usually specify thresholds and are separated by exclamation points, so there are actually two arguments here. The first, 500.0,20%, says that Nagios should generate a warning[1] if the ping packets round-trip time is greater than 500 milliseconds, or if the packet loss is greater than 20%. I know this because I ran the command /usr/local/nagios/libexec/check_ping –h on the command line and read the helpful syntax description.

Port Queries

After ICMP echoes, port queries are the next most common service check performed by most monitoring servers. These plug-ins attempt to open a TCP or UDP connection to a given port on a given host. Port queries are performed by either the check_tcp or check_udp plug-ins. The command definition looks like the one in Listing 6.4.

Listing 6.4   *The Generic check_tcp Definition*

```
define command{
    command_name    check_tcp
    command_line    $USER1$/check_tcp -H $HOSTADDRESS$ -p $ARG1$
}
```

Unlike ping commands, where a warning may be generated, the result of a port query is usually an XOR. Either the port is available or it isn't. Because there isn't much room for ambiguity, no -w or -c is necessary. All that needs to be specified are the hostname and port number. The "all or nothing" nature of the port check affects the notification options, as well. There is no point in specifying that Nagios should send notifications on warnings when warnings cannot occur. Accordingly, the accompanying service description in Listing 6.5 is a bit different from the check_ping service definition in Listing 6.2, in that it specifies different notification options and contains no threshold options.

Listing 6.5   *Check_http service Definition*

```
define service{
    host_name               webServer
    service_description     check_http
    check_command           check_tcp!80
    notification_options    c,r
    use                     chapter6template
    }
```

Listing 6.8 *A check_tcp Wrapper*

```bash
#!/bin/bash
#call check_tcp once for each port; aggregate the result

HOME='/usr/local/nagios/libexec' #path to the plug-ins
PROTO='Nullz0r' #default protocol to use

# a function for printing the help info
printusage ()
{
echo "this plug-in calls check_tcp once for each port"
echo "usage:"
echo "check_multi_tcp -H host -u|-t -p \"port [port] ...\""
echo "-h : print this message"
echo "-H hostname: The hostname of the box you want to query
      (default localhost)"
echo "-p port number: A space separated list of port numbers"
echo "-t wrap around check_tcp"
echo "-u wrap around check_udp"
exit ${EXITPROB}
}

#parse the arguments
while getopts ":hH:utp:" opt
do
        case $opt in
                h )     printusage;;
                H )     HOST=${OPTARG};;
                p )     PORT=${OPTARG};;
                u )     PROTO='udp';;
                t )     PROTO='tcp';;
                ? )     printusage;;
        esac
done

#sanity check
if echo «${PROTO}» | fgrep -q 'Nullz0r'
then
   echo «ERROR: either -u or -t required»
   echo
   printusage
fi

################## Work starts here ##################

#for each port they give us
for i in 'echo ${PORT}'
do
   #call the real plug-in
   ${HOME}/check_${PROTO} -H ${HOST} -p ${i}>/dev/null
```

```
    #did it exit happy?
    if [ «$?» -ne 0 ]
    then
        #no it's not a happy camper
        echo «port ${PROTO}/$i is not available»
        exit 2
    fi
done

#everything's okay, mon
echo «all ports are open»
exit 0
```

Housekeeping aside, this is a simple shell script. It calls the real check_tcp plug-in on each port you give to it in a list and if any of them are not available, it exits with a critical code. If it gets through all of the ports without any errors, then it exits with the okay code. Shell wrappers are utterly ubiquitous among Nagios administrators. If you can't find the functionality you need in the plug-ins directory, before you reinvent the wheel, first check the "contrib" directory of the plug-ins tarball and then ask yourself whether you might be able to coopt a few existing plug-ins in a wrapper to get what you need.

(More) Complex Service Checks

My examples to this point have probably inspired little more than a yawn from you, so in this section, I'd like to branch out a bit and give you a couple examples of real-world monitoring scenarios. I hope these will give you a better feel for the kinds of things that the plug-ins are capable of.

In my first example, Company B uses a rather unreliable combination of filters to block unwanted email on its public MX. The problem is that its business partner, Company A, seems to be particularly disliked by the filters for some reason, so every few weeks, the filters arbitrarily decide to block all email originating from Company A. Various meetings have taken place about the problem, but the combination of filters is so complex and Company B is so large that Company B just can't seem to get it together. Every time they think the problem is fixed, it happens again—and worse, every time it happens, it takes up to a day to figure out that it's happening because the filters at Company B don't bounce the mail. Instead, they answer 250 NotOK and then silently drop the mail on the floor.[3]

To at least provide timely detection of the problem, the sysadmin at Company A defines a command that uses the check_smtp plug-in to periodically perform an SMTP handshake with Company B's mail server. This definition is Listing 6.9.

Listing 6.9 *A Command to Perform an SMTP Handshake*

```
define command{
    command_name       check_spam_block
    command_line       $USER1$/check_smtp -H $HOSTADDRESS$ \
                       -C 'hello companyA.com' -R '250 OK' \
                       -C 'mail from: <alice@companya.com>'\
                       -R '250 OK' \
                       -C 'rcpt to: <bob@companyb.com>' \
                       -R '250 OK'
    }
```

This works well; if Company B answers anything other than 250 OK to any part of the handshake, the administrators at Company A are immediately notified. Further, there's no reason this definition couldn't be expanded to include the data portion of the SMTP conversation, if required.

For the record, you should get permission from someone before you undertake to monitor something that you don't own. This sort of thing can get you into trouble. Another thing to keep in mind is that service checks that actually interact with the services they are watching affect things like logs and connection statistics. If the data portion were included, Bob at Company B would get an email message, and it's always advisable to stop short of doing something that directly affects a human being. On the other hand, poorly written daemons might have problems with service checks that sever the connection at unexpected times. Finally, administrators on the other end might use filters to block access to your monitoring tools if they think the traffic might be malicious in nature. As I said in Chapter 1, "Best Practices," always put some thought into the impact you introduce on the things you monitor, especially if those things don't belong to your company or group.

My second example centers on Ted, who is a systems administrator for a moderately sized health care company. Ted is responsible for obtaining SSL certs from the company's rather shady PKI vendor, VeriSure. Ted is also responsible for registering new domain names, but the company doesn't use VeriSure for this. Recently, Ted's mailbox has been filling up with email from VeriSure. Most of them are marketing mails, offering him discounts to move his company's domain registry to VeriSure.[4] Because his company owns a few domains and SSL certificates, he's getting more than 20 of these per day, so he has a dilemma. Ted wants to /dev/null all email from VeriSure, but he also needs to get SSL expiry notifications. Guess what the command in Listing 6.10 does.

Listing 6.10   *A Solution for Ted*

```
define command{
    command_name    check_ssl
    command_line    $USER1$/check_http $ARG1$ -C 10
    }
```

Check_http is a great plug-in that can do all sorts of useful things. The job of the -C switch is to check the expiry date of a given web site's SSL certificate. If the certificate on the web site expires in fewer than the number of days given (ten, in this case), the plug-in generates a critical error. This solves Ted's problem and is probably a bit more reliable than VeriSure notifications, anyhow.

This definition is the first we've seen that doesn't use the $HOSTADDRESS$ macro. This is because we're specifying a URL, rather than a server address. The URL is passed via an ARG macro, as can be seen in the service description in Listing 6.11.

An interesting digression is that, because the $HOSTADDRESS$ macro is normally the macro that decides which host the plug-in will run on, the host_name directive in the service definition can be whatever you want when that macro isn't used. That is, we can specify some unrelated accounting database server for host_name in Listing 6.11 and the check will work just fine in this example. The only place the host_name directive is used, in the absence of the $HOSTADDRESS$ macro in the command definition, is in the web UI, which will list the check_ssl service as belonging to whatever host_name references.

Listing 6.11   *The check_ssl Service Definition*

```
define service{
    host_name               webServer
    service_description     check_ssl
    check_command           check_ssl!www.myweb.org
    notification_options    c,w,r
    use                     chapter6template
    }
```

E2E Monitoring with WebInject and Cucumber-Nagios

I hope the last two examples made you excited about the types of solutions you can build with the built-in plug-ins. Now it's time to branch out a bit further with an example of "end to end" monitoring. Currently, end to end, or e2e, is all the rage with the monitoring systems vendors. As discussed in Chapter 1, in the section titled "Watching Ports Versus Watching Applications," e2e means that the monitoring system makes use of the service in question in the same way that a user might. This means different things in different contexts. For example, instead of the classic methodology of monitoring port 25 for SMTP, an e2e system would attempt to send an email to itself through the mail system.

WebInject

Because I've been on a bit of an HTTP tangent with these examples, I'm going to continue down that path with one of my very favorite Nagios plug-ins, WebInject. WebInject is a Perl program for performing web site regression testing. With WebInject, you create "test cases" in XML, which describe a list of sites to visit. When visiting each site in turn, WebInject can do lots of useful things, like parsing out and saving strings for later use and verifying the presence or absence of particular text. Perhaps the coolest thing about WebInject is the way it seamlessly handles session states and authentication. For example, WebInject handles cookies automatically in the same manner as your web browser. It saves cookies received from each test case in turn and presents them in the HTTP header of the following test case. For embedded session-ID-based (or, if you prefer, cookieless) authentication schemes, WebInject can parse sessionIDs out of the response text or header of a site and provide it in a variable for later use.

Did I say that WebInject's session handling was the coolest thing? I take it back. The coolest thing about WebInject is that it has a Nagios plug-in mode. With a simple configuration parameter in its config.xml, WebInject becomes a Nagios plug-in, carrying out its test cases and returning with a Nagios-compatible exit code and output string. It even provides performance data.

All this combines to make WebInject the perfect end-to-end web site monitoring plug-in. For example, suppose you ran a web site with a search entry field and a database back-end. Here's one you may be familiar with: www.google.com. In this example, we want to be sure that our web site is operating as expected; that is, not only is it alive, but it's also functional.

In Listings 6.4 and 6.5, we saw that we could monitor port 80 with check_tcp, but this wouldn't catch HTTP errors, such as 404. We improved on this in Listings 6.6 and 6.7 by sending an HTTP GET and parsing the response code, and this is undoubtedly better, but an error in the database back-end could still render the web site useless. Although the HTTP server would be running flawlessly, users wouldn't be getting useful information from their queries. In this example, we'll use WebInject to do an end-to-end check of www.google.com. Our new check_http service will basically reinvent the feeling lucky button. It will go to the web site, perform a query, parse the query results to find the first link, proceed to that link, and verify that the search text appears there. If we can do all that, we know that not only do we have a working web daemon, but also a working web site.

The first step is installing WebInject. It can be obtained from www.webinject.org. Because it's a Perl script, there's not much installing that needs to happen beyond acquiring the various modules it depends upon. By default, WebInject will look for a config.xml file in the present working directory.

As you can see in Listing 6.12, there isn't much to the config.xml I used in this example. The reporttype directive specifies what kind of output WebInject will provide. I set it to nagios, so that WebInject would operate in its Nagios plug-in mode. The testcasefile directive specifies the location of the XML file containing the test cases. Finally, the globalhttplog directive enables logging for the tests. By default, logging information goes into http.log in the present working directory. You may set this option to yes, to log everything, but I've set it to onfail, so that it logs only when there are failures in the test.

Listing 6.12   *The config.xml for WebInject*

```
<reporttype>nagios</reporttype>
<testcasefile>testcases.xml</testcasefile>
<globalhttplog>onfail</globalhttplog>
```

There are two test cases in the testcases.xml file in Listing 6.13. The first test case performs a Google search for the word foo and parses the output to find the first link in the list. The parsing syntax, specified by the parseresponse directive, is unusual but adequate for the task. It uses a single pipe character (|) to separate everything before the desired text from everything after the desired text. So to capture the word bar in the string foo bar biz, the parseresponse syntax would be foo | biz.

After the URL of the first hit has been extracted by parseresponse, it is placed in the WebInject variable {PARSEDRESULT}. We can then refer to this variable in test case 2, which we do in the URL directive. Test case 2 will then proceed to the web site pointed to by the results of test case1 and verify that the word foo exists on that site with the verifypositive directive. Verifypositive supports regex syntax, but we won't need that for this simple example.

Listing 6.13   *The Test Case File for WebInject*

```
<testcases repeat="1">

<case
    id="1"
    description1="goto google. search for foo."
    method="get"
    url=http://www.google.com/search?hl=en&q=foo&btnG=Google+Search
    parseresponse='\<a class=l href="|"'
/>
```

```
<case
    id="2"
    description1="goto {PARSEDRESULT} check for foo"
    method="get"
    url="{PARSEDRESULT}"
    verifypositive='foo'
/>

</testcases>
```

We can test our plug-in by calling webinject.pl from the command line. If everything goes according to plan, you should get some output like this:

```
WebInject OK - All tests passed successfully in 0.704 seconds
|time=0.704;;;0
```

If you delete the reporttype directive from the config.xml, WebInject provides more verbose output, such as that shown in Listing 6.14. It turns out that the first match from Google was the Official Foo Fighters web site. This can be helpful for debugging or for the intellectually curious.

Listing 6.14   *Verbose Output from WebInject*

```
Starting WebInject Engine[el]

-------------------------------------------------------
Test:  testcases.xml - 1
goto google. search for foo
Passed HTTP Response Code Verification (not in error range)
TEST CASE PASSED
Response Time = 0.372 sec
-------------------------------------------------------
Test:  testcases.xml - 2
goto http://www.foofighters.com/ make sure it says foo there
Verify : "foo"
Passed Positive Verification
Passed HTTP Response Code Verification (not in error range)
TEST CASE PASSED
Responsc Time = 0.267 sec
-------------------------------------------------------

Start Time: Sat Aug 12 21:49:57 2006
Total Run Time: 0.718 seconds
```

```
Test Cases Run: 2
Test Cases Passed: 2
Test Cases Failed: 0
Verifications Passed: 3
Verifications Failed: 0
```

All that's left to do now is to create command and service definitions, such as those in Listings 6.15 and 6.16, which are pretty humdrum. Because the query logic is entirely specified within the testcases.xml, there isn't a whole lot left to define in the command and service definitions. It is possible to give WebInject command-line arguments, telling it where to find alternate config files, but the test cases themselves must be read from files. If you thought to yourself that you could write a shell wrapper to create testcase.xmls on-the-fly from Nagios macros, you're well on your way to becoming a Nagios administrator.

Listing 6.15   *A WebInject Command Definition*

```
define command{
    command_name    check_google
    command_line    $USER1$/webinject.pl
    }
```

Listing 6.16   *A WebInject Service Definition*

```
define service{
    host_name               webServer
    service_description     check_google
    check_command           check_google
    notification_options    c,w,r
    use                     chapter6template
    }
```

Cucumber-Nagios

If you liked WebInject's functionality but found its XML test case files offputting, and especially if you're familiar with the Ruby programming language, you should take a look at the cucumber-nagios plug-in.

Cucumber-Nagios, written by Lindsay Holmwood (http://auxesis.github.com/cucumber-nagios/), is an ingenious bit of glue-code that brings together Nagios, Webrat, and Cucumber. Webrat is a Ruby library for testing web applications. It is a browser-simulator API that Ruby programmers can work with to test web sites. Cucumber-Nagios relies on Webrat to run the same sorts of tests Webinject does, but you don't need to interact with it directly to use the plug-in, because the test case files use Cucumber to insulate you from the messy details.

Cucumber is what is known as a "business-readable, domain-specific language." There are many, what my father would call "highfalutin," concepts swimming around in that description that I do not have the space to unravel here, but the point is that Cucumber attempts to make it easy for programmers to describe, in a syntax very close to plain English, how a software feature should work. In accomplishing the goal of describing how a software feature works, Cucumber makes it possible to test that the feature is working correctly, and testing features is really the meat of what it exists to do. Take a look at the example feature file in Listing 6.17.

Listing 6.17   *A Cucumber Feature File*

```
Feature: google search
  Everything between here and "Scenario" is free-form
  description text. It will be ignored by the parser

  Scenario: Search for foo
    When I go to http://www.google.com/
    And I fill in "q" with "foo"
    And I press "I'm Feeling Lucky"
    Then I should see "Foobar"
      But I should not "foo fighters"
```

The language in use in the listing is called "Gherkin." In Gherkin, each software feature is defined in a single file, which carries a .feature suffix. Each line in the file is a statement, and indentation is used to denote structure (spaces or tabs may be used for indentation). When Cucumber parses a feature file like the one in Listing 6.17, it divides the file into features, scenarios, and steps. It then looks for the keywords (Given, When, Then, And, But) that begin each step and attempts to match the portion of each line after the keyword to a predefined Ruby code block called a "step definition."

Listing 6.18 shows the step definition that matches the line:

```
And I fill in "q" with "foo"
```

Listing 6.18   *Step Definition Example*

```
When /^I fill in "(.*)" with "(.*)"$/ do |field, value|
  fill_in(field, :with => value)
end
```

Even if you're not familiar with Ruby, it's pretty easy to recognize that a simple regular expression is being used to match the statement and capture the form field name and value into two variables, which are then passed to the "fill_in()" function (a function provided by Webrat). The magic of cucumber-nagios lies mostly in the numerous step definitions that have been provided. Pretty much every definition you might need to perform complex checks on web applications, HTTP Headers (and therefore web-services interactions), ICMP and DNS services, and a few other odds and ends are included.

With just a little bit of Ruby knowledge, the cucumber-nagios framework is easy to extend, and with Ruby's vast array of third-party libraries, any sort of end-to-end service check, such as the email system checks I spoke about earlier, could be implemented within it.

To install cucumber-nagios, use Ruby's "gem" tool:

```
gem install cucumber-nagios
```

After it is installed, you can use the "cucumber-nagios-gen" tool to generate a top-level directory that contains everything cucumber-nagios needs to run your service checks:

```
cucumber-nagios-gen project myCucumberServicechecks
```

There will now be a directory called "myCucumberServicechecks" in the present working directory, wherein you will find a directory called "features," which contains a directory called "steps." The predefined step definitions can be found there, and reading through these files is the only reliable way to find out what steps you may use to build your tests. After you have a project directory, the gen tool can help you generate feature files for it:

```
cucumber-nagios-gen feature myCucumberServicechecks google.com
```

You may now edit features/myCucumberServicechecks/google.com.feature to define your test. When you're ready to try it out, simply run the following:

```
cucumber-nagios features/myCucumberServicechecks/google.com.feature
```

The cucumber-nagios plug-in will return Nagios-compatible output, perfect for codifying as a service check in your services.cfg.

I hope this first section provided you with a good idea of the remote querying capability of Nagios. Now it's time to move on to remote execution.

Watching Windows

Windows can be a challenge for an administrator building monitoring systems using Nagios or otherwise because it is a bit more of a black box than most UNIX environments. NSClient, a Windows-specific plug-in for Nagios, provides just about all the functionality one could want, but even it presupposes some knowledge about the current Microsoft scripting environment, especially WMI. In fact, making heads or tails of the Windows scripting environment is probably the largest barrier to entry for someone who wants to monitor Windows, so let's tackle that first. If you're already adept at programming and scripting in Windows, feel free to skip ahead to the subsection titled "Getting Down to Business."

The Windows Scripting Environment

Google "Microsoft Scripting" and you'll get back a dizzying array of acronyms and product names: WSH, OLE, Cscript, WMI, ADSI, JScript, VBScript, and PowerShell, to name just a few. If you're wondering, "Whatever happened to batch?" this section is for you.

Beginning with DOS and OS/2, batch scripts were used to automate tasks. These scripts were little more than lists of DOS commands in a file, which the command.com program could execute. Although they possessed rudimentary functionality and cumbersome syntax, batch scripts managed to scale well for many systems administration tasks.[5]

However, something more robust was needed, so about the time Windows 98 was introduced, so was WSH. WSH, or Windows Script Host, is a language-independent execution environment for scripts. For most purposes, you can think of it in the same terms as any interpreter you are familiar with, such as Perl or Python. What makes WSH different is its use of modular engines to provide syntax, so although you speak Perl to the Perl interpreter and Python to the Python interpreter, WSH has no native syntax. In fact, it's possible to speak both Perl and Python to WSH in the same script. WSH provides the execution environment, some common data structures and I/O hooks, and leaves the specific syntax up to the language engine.

WSH, as installed by Microsoft, includes only language engines for VBScript and Jscript, and in practice, most people use the VBScript syntax. Because of this, many people refer to scripts executed by WSH in general as VBScripts, or Visual Basic Scripts.

VBScript, or Microsoft Visual Basic, Script Edition, is a subset of Microsoft Visual Basic programming language, which, in turn, owes its lineage to the original Beginner's All-Purpose Symbolic Instruction Code developed at Dartmouth College. The vbscript.dll script engine interprets code written in VBScript, and the script engine can be used by either the ASP engine in Internet Explorer for web applications or WSH for systems programming and automation. So VBScript wears two hats: It is both a web application language and a general purpose scripting language. In its web application role, VBScript is embedded into HTML (similar to JavaScript) to be interpreted by the web browser. In its standalone or WSH role, VBScript is executed by WSH from a file, usually possessing a .vbs extension.[6]

To muddle things further, WSH is composed of two separate execution environments: Wscript and Cscript. These environments, implemented as two separate programs, are identical except that Wscript is GUI-based and Cscript is command-line driven. For example, the following snippet:

```
Wscript.Echo "Hello World"
```

when executed by Wscript, opens a new window that contains the words "Hello World." The same code executed by Cscript simply prints "Hello World" to stdout at the command prompt.

WSH scripts are self-contained files with extensions that denote their syntax. VBScript WSH scripts usually end in .vbs and can be executed with either Cscript or Wscript in the following manner:

```
cscript foo.vbs
wscript foo.vbs
```

When the execution environment (or host, in Microsoft parlance) is not specified, the default host is chosen. The default host, out of the box, is (of course) Wscript, which is probably not what you want. A few other annoyances are built into WSH, such as its habit to output a banner that informs you that your script was run by Microsoft's WSH. In Cscript, this banner is actually injected into the output of your program (STDOUT, not STDERR). This is bad if you're writing Nagios plug-ins, because Nagios will parse only the first line of text output by the plug-in. Switches exist to change the default behavior. The following is what most people use:

```
cscript //I //nologo //H:cscript //S
```

The I switch specifies interactive mode, rather than batch mode. Nologo removes the banner. The H switch specifies that Cscript should be the default script host, making it possible to launch foo.vbs without first specifying Cscript. Finally, the S switch saves these settings to the Registry, making them permanent.

COM and OLE

Whereas UNIX relies on small, text-based, single-purpose programs that work together toward accomplishing tasks, Windows, as an environment, tends toward large monolithic graphical programs. This poses a dilemma to would-be automators: How do you script a GUI? Enter COM. Since 1993, the Component Object Model and related technology have attempted to provide a language-agnostic interface to software that is otherwise immune to automated integration.

Software developers using COM build their applications using COM-aware components. If implemented correctly, these components provide interfaces to the applications' functionality via any other program that speaks COM. These interfaces can be used for any number of purposes, such as interprocess communication or even automation. OLE is COM's object model. OLE gives COM objects their names and specifies things such as object inheritance. Most people associate OLE with embedding an Excel spreadsheet within a word document,[7] but it is now much more powerful than that.

Because most important applications in Windows expose their functionality via COM, and WSH provides access to any COM object, it is possible to use scripting languages such as Perl and VBScript to automate applications in Windows, including everything from programmatically creating Excel documents to driving MMCs (Microsoft Management Consoles). OLE and COM provide the glue with which Nagios may be tied to any application that exports its functionality via COM, which is to say most applications out there. Scripts that use COM to query information from Windows systems and then exit with the appropriate exit codes are, by definition, Nagios plug-ins.

WMI

One piece of software that doesn't export its functionality via COM is the kernel. Various flavors of UNIX have their proc or sysfs filesystems, but until recently, this critically important system information was largely unavailable to scripting languages in Windows. The closest thing was perfmon, which is a real-time performance statistics program that didn't lend itself to being driven from scripting languages or the command line. Products such as SNMP Informant could export perfmon information via SNMP, but this is more kludge than solution. Windows Management Instrumentation fills this gap nicely. WMI is like a COM interface to the runtime environment. It can be thought of as another COM interface, but one that provides access to things such as disk and network utilization.

WMI is derived from the Distributed Management Taskforce's CIM concept. CIM, or Common Information Model, is a large database (called a CIM Repository) of objects that represent manageable entities, such as hard drives, entire computer systems, and software packages. WMI is Microsoft's implementation of CIM. The CIM concept encompasses more than what you'd find in /proc. It is a collection of information about computer systems that includes the current memory utilization, as well as things such as the system's serial number and PowerPoint version. As such, many current Windows applications extend the CIM database with their own information. Applications and drivers that provide information to the CIM are called providers.

Because the CIM repository contains a lot of information, it is broken down into namespaces specific to provider types, such as Root\SNMP or Root\MicrosoftIISv2. The built-in OS providers use Root\CIMv2. These namespaces are further broken down into classes, which are functionality specific objects, such as Win32-Process or DiskObject. Additionally, WMI implements an SQL-like query language called WQL to help you find the specific pieces of information you are looking for. Very few people programmatically explore the CIM Repository using WQL, however. Most use a GUI browser, such as wbemtest.exe, located in system32/wbem on most Windows systems. WQL is still necessary, however, to instantiate specific class objects in a script.

To give you a feel for the capabilities of WMI/OLE, as well as what a Nagios plug-in that uses WMI looks like, I've included Listing 6.19, which is a Nagios plug-in called check_ dllHost. Its purpose is to make sure that no single dllHost process consumes more than a user-specified amount of RAM. It is written in Perl and uses the Win32_Process WMI Class.

Listing 6.19 *Check_dllHost*

```perl
#!/usr/bin/perl
#a plug-in to check whether any dllHost processes are
#eating too much ram
#Blame Dave Josephsen --> Wed Apr 20 13:23:10 CST 2005

#########variable init#########
#we use the win32::ole module to connect to wmi
use Win32::OLE;
use Win32::OLE qw (in);

#our warning and critical thresholds are passed via arguments
$warn=$ARGV[0];
$warn='200000000' unless ($warn);
```

```
$crit=$ARGV[1];
$crit='250000000' unless ($crit);

$counter=0;

#########real work begins here#########
#spawn a wmi object
$oWmi = Win32::OLE->GetObject("WinMgmts://./root/cimv2")
        or die "no wmi object";

#this wql query gets all the processes running on the box
$oProcessEnumObj=$oWmi->
                ExecQuery("Select * from Win32_Process ");

#iterate through the process list. Retrieve  'dllHost' procs
foreach $oProcess ( in($oProcessEnumObj) ){

        if($oProcess->Name =~/dllHost.*/i){
          $counter += 1 ; #keep track of how many there are for later

      #are you using up my ram?
          if ( $oProcess->WorkingSetSize >= $crit){

          #you sure are
              print "CRITICAL ". $oProcess->WorkingSetSize .
                    "kb in use by ". $oProcess->Name . "\n";
        exit 2;
         } elsif ( $oProcess->WorkingSetSize >= $warn){
             print "WARNING ". $oProcess->WorkingSetSize .
                    "kb in use by ". $oProcess->Name . "\n";
        exit 1;
          }
      }
}
#if we made it this far, then everything's all right, mon
if($counter >= 0){

   print "OK ". $counter . " dllHosts running,
        none over the limit \n";
   exit 0;
}else{
   print "OK no dllHost processes running\n";
   exit 0;
}
```

Scripts that use WMI almost invariably follow the same pattern. Spawn a WMI object, use WQL to query some subset of information from the object, check the status of that information against thresholds or expected results, and exit. The WMI URL[8] ("WinMgmts://./ root/cimv2") is very important. It specifies from where in the CIM our WMI object is derived, which limits the kind of information we can use WQL to query. Also important is the WQL query:

```
$oProcessEnumObj=$oWmi->ExecQuery("Select * from Win32_Process ");
```

The syntax "Select * from Win32_Process" is WQL. Its purpose is self-explanatory: it returns all the currently running processes. The result of a WQL query is always a collection object. A collection object is a fancy sort of array that doesn't behave in the usual Perl manner. To ease iteration across collection objects, the Win32::OLE module provides the "in" function. The "in" function makes possible constructs such as this:

```
foreach $oProcess ( in($oProcessEnumObj) ){
```

and this:

```
@oProcesses=in($oProcessEnumObj)
```

possible. Python treats collection objects as enumerations, so the usual "for...in" will work.

The dot (//./) in the URI specifies that the WMI object in question is spawned on the local system. If we replaced the dot with the hostname of a remote host, we could consume WMI information from a remote host via RPC over the network.[9]

Listing 6.20 is another WMI/Perl Nagios plug-in. Although its purpose—to determine if any services in a cluster are not currently online—is very different from Listing 6.19, the pattern is nearly identical.

Listing 6.20   *A check_cluster Plug-in in Perl/WMI*
_____

```perl
#!/usr/bin/perl
#check_cluster a perl script/nagios plug-in to check if any cluster
#resources are in a state other than online.
#blame Dave Josephsen --> Sat Jan 22 17:03:34 CST 2005

#########variable init#########
use Win32;
use Win32::OLE qw (in);

#swap these if you want to take the servername as an arg
$server=".";
#$server="$ARGV[0]";

#unlike most WMI classes, MSCluster is derived from
#/Root/MSCluster, so we have to specify that
#in the class path.
$class = "WinMgmts://"."$server"."/Root/MSCluster";
```

```perl
#the MSCluster provider classes are just barely documented here:
#http://msdn.microsoft.com/library/default.asp?url=/library/en-us/
mscs/mscs/server_cluster_provider_reference.asp
$object_class='MSCluster_Resource';

#possible resource states, from;
#http://msdn.microsoft.com/library/default.asp?url=/library/en-us/
mscs/mscs/clusresource_state.asp
$state{-1}='Unknown';
$state{0}='Inherited';
$state{1}='Initializing';
$state{2}='Online';
$state{3}='Offline';
$state{4}='Failed';
$state{128}='Pending';
$state{129}='Online_Pending';
$state{139}='Offline_Pending';


#########real work begins here#########

#get a wmi object
$wmi = Win32::OLE->GetObject ($class);

#get a collection of resources off the cluster
$resource_collection=$wmi->
                    ExecQuery("Select * FROM $object_class");

#how many resources are there?
$max=$resource_collection->{Count};

#the 'in' function comes from Win32::OLE
#it's the same thing as: Win32::OLE::Enum->All(foo)
@collection=in($resource_collection);

#are any resources in any state other than online?
for ($i=0;$i<$max; ++$i){
   if($collection[$i]->{State}!='2'){
      push(@broken,$collection[$i]);
   }
}

#if so, do bad things
if(scalar(@broken)>0){
   foreach $j (@broken){
      print("$j->{Name} is $state{$j->{State}}, ");
   }
   exit(2);
}else{
   #otherwise, do good things
   print "$max resources online\n";
   exit(0);
}
```

To WSH or Not to WSH

Although the scripts in my examples run on Windows and consume information from WMI objects, they don't, in fact, use WSH. If you choose to script in any language other than VBScript, you'll have to install the language engine, which means you'll also have a choice of interpreters. ActiveState Perl, for example, installs both the Perl interpreter and the Perl WSH language engine. Scripts with Perl syntax, executed via Cscript and possessing a .pls extension, are executed by WSH, whereas those with a .pl and called without a script host are executed by the native Perl interpreter, perl.exe.

Within WSH, things are a bit different. For example, you can't use modules, and Perl switches, such as -w, don't work. The ARGV array is absent, as well, because perl.exe, in one way or another, provides all these things. The WSH environment instead provides a $Wscript object with which you can parse arguments and make connections to OLE objects. In WSH, $Wscript even supersedes print(), so that

```
print("foo $bar");
```

becomes

```
$Wscript->Echo("foo",$bar);
```

In Listing 6.19, I used the Win32::OLE CPAN module to connect to WMI via COM. Within WSH, this becomes a little easier.

```
use Win32::OLE;
use Win32::OLE qw (in);
$oWmi = Win32::OLE->GetObject("WinMgmts://./root/cimv2")
        or die "no wmi object";
```

becomes

```
$oWmi = $Wscript->GetObject("WinMgmts://./root/cimv2");
```

The choice is yours. WSH doesn't provide anything that native Perl with Win32::OLE doesn't, and if you're adept at Perl, you would probably rather use @ARGV than $Wscript->Arguments. The situation is the same for Python and Ruby. If alternative language engines were installed out of the box, WSH might be a compelling alternative. As it is, most people who don't use VBScript use the native interpreter for their language of choice. For more information on programming with WSH in Perl, Python, Ruby, or even Object REXX, check out Brian Knittel's book, *Windows XP, Under the Hood*.

To VB or Not to VB

The fact that the VBScript language engine is installed by default on all current versions of Microsoft Windows makes it the prevalent scripting language for the platform. The extent to which this is true is difficult to express. VBScript is so popular that it's nearly impossible to find a problem for which three or four scripts have not already been posted on the Internet. So prevalent is sample code that it's not even necessary to know the language at this point. In fact, I've yet to actually meet a Windows administrator who knows the language well enough to write a script from scratch.[10] It's as if every Visual Basic Script currently in existence is a modification of a modification of some original Visual Basic Script that was written from scratch 12 years ago. Microsoft technet boasts a tool they are particularly fond of, called the scriptomatic, that generates VBScript for you from pull-down menus. Microsoft-hosted courses on the subject of automation, at this point, only bother to teach enough of the syntax to enable students to download and modify existing code.[11]

I digress. Getting back to the point, VBScript is a supportable choice for system automation and monitoring on Microsoft Windows. If you are familiar with Java, you can probably pick up the syntax in a weekend, but its only real advantage is that it is available on all Windows servers out of the box. Its lack of case sensitivity, unwieldy regex syntax, and strange habit of using list context in inappropriate situations make it unpopular among people who actually enjoy programming, but administrators who might otherwise use another language choose VBScript so they don't have to install an interpreter on every machine.

For that reason, I want to mention that tools exist for most of the popular scripting languages that make it possible to package scripts in a standalone manner. PAR for Perl, py2exe for Python, and rubyscript2exe for Ruby create .exe programs out of scripts. The exe's can then be run on Windows machines without the interpreter. The binaries created by these tools can be quite large[12] and they don't run any faster in their compiled form than they do in the interpreter, but they enable you to write scripts and Nagios plug-ins in the language of your choice and run them on any Windows machine in your environment, without having to install the interpreter on each Windows host.

The Future of Windows Scripting

Currently, in the world of Microsoft, .NET is all the rage. The .NET framework is a large class library of solutions to commonly occurring programming problems. It is meant to eventually supercede many existing Microsoft technologies, including COM and OLE. There is no plan, however, to stop development or support of COM at Microsoft, and .NET does little to address the needs of systems people, so I suspect the VBScript/COM/OLE combo to remain the systems scripting environment of choice for the foreseeable future.

PowerShell, formerly known as MSH, or Mondad, will probably have a larger impact on scripting on the platform going forward than .NET. PowerShell is a scriptable CLI that is currently available as a download. PowerShell is capable of doing anything WSH can do, while providing a much friendlier interface to systems administrators. Where COM and VBScript provide a scriptable interface to existing GUI tools that support COM, Microsoft claims that future versions of system configuration utilities will be written in PowerShell with GUI wrappers, thus ensuring a scriptable interface to the OS going forward.

While the PowerShell implementation borrows concepts from UNIX, such as passing information between small distinct programs via pipes, PowerShell does this in a much more object-oriented manner. For example, PowerShell programs (called cmdlets) give output to the shell in text, but when the same output is piped to another PowerShell cmdlet, data objects are exchanged instead. Microsoft is fond of saying that this completely eliminates the need for text processing tools such as awk, sed, and grep. To see what they mean by that, consider the list of servers in Listing 6.21.

Listing 6.21   *A List of Boxes*

```
Name,Department,IP
frogslime,R&D,12.4.4.17
tequilasunrise,Finance,12.4.5.23
151&coke,R&D,12.4.4.151
theangrygerman,MIS,12.0.0.2
7&7,Finance,12.4.5.77
```

Let's say you want to extract and print only the machines in the finance department. In UNIX, you'd probably do something such as the following:

```
cat list.csv | grep Finance
```

Or, if you wanted to be specific:

```
cat list.csv | awk -F, '$2 ~ /Finance/ {print}'
```

In these examples, grep and awk are given the comma-separated list as a text input stream, and they filter their output of the text according to the options we give them. Grep was told to print the lines containing "Finance," and awk was told to print the lines with "Finance" in the second field, where fields were separated by a comma.

PowerShell cmdlets don't receive text input from each other, so although the semantics of the following line of code is similar to the UNIX examples, it operates in an altogether different manner:

```
Import-csv list.csv | Where-Object {$_.department -eq "Finance"}
```

The PowerShell cmdlet Import-csv reads in the contents of list.csv and creates an object of type csvlist.[13] This csvlist object is then passed to the Where-Object cmdlet, which uses the department property of the object to extract the records matching "Finance." Because the pipeline ends here, the Where-Object cmdlet will dump its findings in plain text to the shell. If the pipeline had continued, however, the Where-Object cmdlet would have provided the modified csvlist object to the next cmdlet in line. The Perl-like $_ object (which represents the current iterator), curly braces, and pipes combine to give the command a very UNIX-ish feel. The syntax is far more appealing to systems-type scripting people than VBScript, which is why I think PowerShell has a better than average chance of catching on.

Microsoft has also made the statement that the UNIX pipes model has made sysadmins into expert text manipulators,[14] and this is undoubtedly true. To that point, I would add that along with being expert text manipulators comes the expectation that data will be manipulated and formatted in the manner specified by the person performing the manipulation. Given this, I think that many administrators may hear PowerShell's message as, "Give us your data and we'll tell you what you can do with it." So I see where Microsoft is going when they say that the object model will eliminate the need for text processing, but I humbly predict that the most popular PowerShell site on the Internet, besides Microsoft's, will be the site providing text processing cmdlets for use alongside the official Microsoft ones.

I think PowerShell is a very promising technology whose time is far overdue, and although I'm skeptical that the object model makes a compelling case for the wholesale abandonment of text-processing tools, I'm glad that Microsoft is headed in this direction. PowerShell is certainly a net gain, and I expect that it will greatly expand the use of scripting and automation among systems professionals on the Windows platform. Although the current VBScript/WSH/COM combination will probably remain the de facto scripting standard for a while yet, because of its overwhelming mindshare and default availability, there's no reason you couldn't use PowerShell to get work done right now.

Getting Down to Business

Now that we have a pretty good understanding of the environment and where the good information is, let's check out some ways to extract the stats we need and provide them to Nagios. There are three popular ways to glue Windows plug-ins to Nagios: NRPE-NT, check_nt, and NSClient++. Of the three, NSClient++ (also referred to as NSCP), is the only tool being actively developed as I write this. The other two projects are still usable, but abandoned. I include them here for reference only.

NRPE

NRPE, as described in Chapter 2 and Chapter 3, "Installing Nagios," is a lightweight client and server that enables the Nagios server to remotely execute plug-ins stored on the monitored hosts. NRPE is great when you have existing monitoring scripts that you want to use with Nagios or when you have a need for custom monitoring logic. Any scripts you may have written (or downloaded) to monitor your Windows boxes can be defined as a service in Nagios and called via NRPE-NT, the Windows port of NRPE.

For example, to run the check_dllhost plug-in from Listing 6.19 with NRPE-NT, first obtain and install NRPE-NT, as described in Chapter 3, in the section "Installing NRPE." After the NRPE service is installed on the Windows host, add the following line to the bottom of the nrpe.cfg file:

```
Command[check_dllhost]=c:\path\to\check_dllhost.pl 200000000
250000000
```

This line tells the NRPE daemon on the Windows host what to do if someone asks it to run check_dllhost. The Nagios server must then be configured to run the check. The first step is to define the command in the commands.cfg, as in Listing 6.22.

Listing 6.22   *Check_dllhost Command Definition*

```
define command{
    command_name    check_dllhost
    command_line    $USER1$/check_nrpe -H $HOSTADDRESS$ \
                    -c check_dllhost
    }
```

The command definition in Listing 6.22 makes use of the check_nrpe plug-in. As mentioned in previous descriptions of NRPE, check_nrpe is the client portion of the NRPE package. In the preceding definition, it is given two arguments: -H, with the hostname of the Windows server, and -c, with the name of the command to be executed on the remote server. After the command has been defined, Nagios can schedule it for execution. This is done with a service check command, so the next step is to define the service, such as the one in Listing 6.23.

Listing 6.23   *Check_dllhost Service Definition*

```
define service{
    host_name               windowsServer
    service_description     check_dllhost
    use                     chapter6template
    notification_options    w,c,r
    }
```

After this is done and the Nagios daemon is restarted, Nagios will poll the check_dllhost program on a regular basis. NRPE may be used to execute any script on any Windows host in this manner.

So now that you know how to schedule plug-ins, the question becomes what to schedule. One popular answer to this question is the basic NRPE plug-ins for Windows, available at the NagiosExchange here:

```
www.nagiosexchange.org/NRPE_Plugins.66.0.html?&tx_netnagext_pi1[p_
➥view]=62
```

This package includes plug-ins to check the usual metrics: CPU, memory utilization, and disk space. It also contains a plug-in to query the state of arbitrary services and an eventlog parser. The basic plug-ins are DOS programs written in C and satisfy the basic monitoring needs most people have. This has the added bonus of being the closest, methodologically, to remote execution in UNIX, so if you like standardized methodologies, this may be for you.

Check_NT

Another popular Windows plug-in is Check_NT. Check_NT is a monolithic standalone Nagios plug-in that comes with its own client. In other words, Check_NT is not executed by way of NRPE, but by its own client software on the Nagios server, via its own network protocol. NSClient contains all the functionality mentioned previously, plus the capability to query any perfmon counter, which makes it an attractive alternative to NRPE with the basic pack for people who don't have any custom plug-ins. Running check-nt -h on the Nagios server will return a rundown of its features and use.

Check_NT is still the plug-in used by the majority of people who watch Windows hosts with Nagios. It's very simple to install and contains just about all the functionality that most people want starting out. Check_NTcannot run external scripts; it provides only the functionality built into it. This means that if you want to run check_nt and a few custom scripts, such as the ones in Listings 6.19 and 6.20, you'll need to either install two services on each Windows host: the Check_NT service and NRPE, or go with the third option, NSClient++.

NSCP

NSClient++ aims to replace both Check_NT and NRPE by combining their functionality into a single program. It contains all the query functionality of Check_NT, plus a built-in implementation of the NRPE protocol. NSClient++ will listen on both the NRPE and check_nt TCP ports, and respond to queries just like the clients for those tools expect, so it is, in effect a drop-in replacement for both tools. Additionally, NSClient++ supports WMI query functionality and will execute external scripts as well as Lua and Python scripts internally.

The NSC++ command in the definition in Listing 6.24 will check the CPU load of a Windows box using the check_nt protocol.

Listing 6.24   *Check_nt_cpuload Command Definition*

```
define command{
    command_name     check_nt_cpuload
    command_line     $USER1$/check_nt_wrapper -H $HOSTADDRESS$ \
                     -p 4242 -v CPULOAD -l $ARG1$
}
```

In the preceding command example, the -p switch specifies an alternate listening port. The default NSClient port (1284) conflicts with Microsoft Exchange RPC, so you might want to change it. The -v switch specifies which variable to check. Valid variables include the following: CLIENTVERSION, UPTIME, USEDDISKSPACE, MEMUSE, SERVICESTATE, PROCSTATE, and COUNTER. Each type of variable may take options. These options are specified by the -l. The CPULOAD variable takes a comma-separated triplet in the form <minutes range>,<warning threshold>,<critical threshold>. If I wanted to take a five-minute CPU average and warn on 80 and alarm on 90, I would specify 5,80,90. Multiple triplets can be specified, so if I wanted a five-minute average and a ten-minute average with slightly lower thresholds, I could use 5,80,90,10,70,80. The command definition in Listing 6.24 uses the triplet specified by argument macros in the service definition in Listing 6.25.

Listing 6.25   *Check_nt_cpuload Service Definition*

```
define service{
    host_name               windowsServer
    service_description     check_nt_cpuload
    check_command           check_nt_cpuload!5,80,90,10,70,80
    use                     chapter6template
    notification_options    w,c,r
    }
```

NSClient++ packs a dizzying array of functionality into a very small package and, going forward, I don't see a compelling reason to use anything else (except for perhaps Check_MK, but more on that later). It is an arguably more complex install than NRPE alone, but the install is well worth the functionality it provides.

Watching UNIX

Compared to Windows, the systems programming and automation scene for UNIX users is pretty straightforward. Most sysadmins tend to use some combination of C, Perl, Python, or shell programming to write automation and glue code and, although these tools continue to evolve and get better, the overall systems programming scene doesn't change a lot from year to year. Further, in contrast to Windows, where using any language other than VBScript means downloading and installing an interpreter, UNIX administrators are far more likely to have more interpreters and compilers installed by default than they have a need for.

Combine a rich assortment of available tools with the fact that Nagios was written to run on Linux and you get a plethora of UNIX-based Nagios plug-ins to monitor all manner of things written in all sorts of languages. In fact, when posting a question, such as "How do I monitor X on my freebsd box?" to the nagios-users list, it's not uncommon for more than half of the responses to be written in code of one type or another. This section should give you a good feel for the plug-ins people frequently use. I'll set up checks for the "big three:" CPU, Disk, and Memory. I'll also cover some of the details of what these metrics mean.

NRPE

By now it should be no surprise that NRPE is the remote execution tool of choice for Nagios plug-ins on UNIX boxes. Unlike Windows, where some multipurpose plug-ins, such as NSClient, have implemented their own daemons and protocols, plug-ins on the UNIX side have stayed single-purpose and assume the use of a remote execution program, such as NRPE or check_ssh. As for the plug-ins, most people start out with the official plug-ins tarball that the Nagios daemon uses.

Different hosts will build different plug-ins from the tarball based on the libraries they have installed. So it's not particularly wasteful to have a single set of plug-ins for both the Nagios daemon and the hosts it monitors. For example, check_snmp will compile only if the host in question has the net-snmp libraries installed. This is true for quite a few of the plug-ins in the tarball. Check out Chapter 3 for details on getting and installing the official plug-ins tarball and NRPE.

CPU

Although measuring CPU utilization may seem a relatively straightforward task at first glance, it is, in fact, an intricate and complex problem with no easy solutions. Two CPU-related metrics are normally used to summarize CPU utilization. The first is the classic percentage-based metric. This is a number representing the percentage of CPU utilization occurring now. For example, "The CPU is currently 42 percent utilized." This number is how most people who aren't UNIX administrators understand things, so let's look at it first.

If you ask embedded systems or computer engineers, they will tell you that processors in the real world are either utilized or not. There is, in fact, no such thing as 42 percent utilization on a microprocessor. At any given instant, the CPU is computing some bit of machine code, or it is idle. So if the only two numbers that exist are 0 percent and 100 percent, what can this percentage number actually mean?

In fact, the CPU is never actually idle; "idle" is just one of many states that the processor spends its time computing. Idle happens to be the lowest priority state, so when the CPU spends its time looping in the idle state, it's still doing work, just low priority, preemptable work.

Therefore, to be meaningful, the utilization percentage must be something akin to an average of processor state versus time. Exactly what is being averaged, and for how long, is a question that is answered in software, so even on the same OS, two separate performance applications can measure it differently. Because all processors have a lowest priority state, a popular methodology for providing a single percentage number is that of measuring the percentage of time that the CPU spends in its lowest priority state and then subtracting this number from 100 to obtain the actual utilization percentage. In other words, the utilization percentage is the percentage of time between two polling intervals that the CPU spends in any state other than idle. There are many other ways. For example, some processors have built-in performance counters that may be queried with system calls, so using these is a popular alternative.

The bottom line is that the classic CPU percentage metric presents all sorts of problems from a monitoring perspective. It is an overly volatile and ambiguous metric that doesn't necessarily reflect the load on a system; therefore, it isn't a good indicator of problems. Even in a capacity-planning context, the number has questionable value. For example, 100 percent CPU utilization can be a good thing if you are trying to optimize system performance for an application that is bandwidth- or Disk I/O-intensive. My advice is to avoid this metric in systems monitoring when you can.

The second metric is that of UNIX load averages. This is a set of three numbers that represent the system load averaged over three time periods: 1, 5, and 15 minutes. You may recognize them from the output of several different shell utilities, including top and uptime. These load average numbers are exponentially damped and computed by the kernel, so they tend to be less volatile than the CPU percentage metric, and they are always computed the same, no matter whom you ask. Exactly what these numbers represent is a question that is difficult to express without using math. In fact, the deeper one delves, the harder the math becomes, until first order derivative calculus becomes involved. If you're into that sort of thing, I'll refer you to Dr. Neil Gunther's paper: www.teamquest.com/resources/gunther/display/5/index.htm.

For the rest of us, I'll attempt an explanation in English. The current load average is the load average from five seconds ago, plus the "run-queue length." The run-queue length is the sum of the number of processes waiting in the run-queue plus the number of processes that are currently executing. To save on kernel overhead, the kernel doesn't actually compute three separate load average numbers. The three numbers in the triplet are computed from a common base number. When computing the load average for each time period in the triplet, a different exponential factor is applied to both the five-second-old average and the current run-queue length to dampen or weigh the values accordingly. The ratio used by the dampening/weighing factor is somewhat controversial, at least in Linux, but the load triplet, in my experience, is a useful metric despite disagreement over exponential dampening ratios.

So the load average metric is directly tied to the number of processes waiting for execution. Each new process waiting for execution adds a "1" to the run-queue length, which affects the load average in the manner described previously. This is a much more practical metric of server utilization because it effectively captures how capable the system is of keeping up with the work it is being given. For a single-CPU system, a load of 1 would effectively be 100% utilization, but a better way of thinking about it is that the system has exactly enough capacity to handle the current load. A load average of .5 would mean that the system has twice the capacity it needs to handle the current load and 3 means the system would need three times the capacity it currently has to handle the load. For multi-CPU systems, the load numbers should be divided by the number of CPUs, so a load average of 3 on a four-CPU system means the system is 75 percent utilized, or it has a quarter more capacity than it needs to handle the current load.

There are two problems with the utilization triplet that you should be aware of. The first, and probably worst of the two, is that the triplet is not understood by laymen, such as managers and execs, and you have no hope of educating them.[15] The second problem with the triplet is that the run-queue length is not strictly CPU-bound. Bad server problems, such as EXT3 panics, can also cause processes to back up in the run-queue, at least on Linux. In practice, this turns out not to be too bad of a problem because anything that is backing up the run-queue is bad news and bears your attention, anyway.

Various pearls of wisdom are floating around in books and on the web, which say things such as: "load averages above 3 are bad." I would agree that you probably want to upgrade a single-CPU system that is perpetually loaded at 3, but in the context of setting monitoring thresholds, nothing is written in stone. In the real world, I've seen systems go as high as 25 before showing any real signs of latency, and I have several boxes I don't worry about until they get at least that high. Figuring out where your thresholds should be is definitely a job for you, so check out my discussion of baselines in Chapter 1, "Best Practices," and decide accordingly.

The first step in setting up a CPU check is to add the check_load command to the nrpe.cfg file on the monitored host. The command at the bottom of the file should look something like this:

```
command[check_load]=/usr/libexec/check_load -w $ARG1$ -c $ARG2$
```

Notice the macros in the nrpe.cfg definition. In previous examples with the nrpe.cfg file, I've used static thresholds. Static thresholds are generally preferable, from a security standpoint, because potentially dangerous data is not accepted by the NRPE daemon. In practice, most people use tcpwrappers to control access to the daemon and use argument passing to centralize the thresholds on the Nagios server. Static thresholds on a per-host basis quickly become too much for most administrators to manage.

To be able to pass arguments, you must first compile NRPE with—enable-command-args, and set dont_blame_nrpe to 1 in the nrpe.cfg.[16] After this is done, you can send arguments to the NRPE daemon via the command definition on the Nagios server. Listing 6.26 has an example command definition for use with an argument-accepting NRPE daemon.

This command definition picks up the thresholds from the service definition via argument macros and passes them on to the NRPE daemon on the remote host. Newer versions of check_nrpe support two syntaxes for argument passing. The classic way is the command name with exclamation-mark-separated arguments, and the new way specifies space-separated arguments after an -a switch. Although the new way is more readable, I prefer the old way because character escaping is cleaner.

Listing 6.26   *Check_load Command Definition with Argument Passing*

```
define command{
        command_name     check_load
      #old way:
      command_line     $USER1$/check_nrpe -H $HOSTADDRESS$ \
                       -c check_load!$ARG1$!$ARG2$
        #new way:
```

```
         #command_line     $USER1$/check_nrpe -H $HOSTADDRESS$ \
         #                 -c check_load -a $ARG1$ $ARG2$
}
```

Listing 6.27 contains the accompanying service definition. This is where the thresholds are specified by the administrator. The check_load plug-in in our example takes two arguments: one for the warning threshold and one for the critical threshold. Thresholds are specified as UNIX load average triplets. In our example, Nagios generates a warning for someUnixHost if its 1-minute average goes above 15, its 5-minute average goes above 10, or its 15-minute average goes above 5.

Listing 6.27 *The check_load Service Definition*

```
define service{
        host_name                 someUnixHost
        service_description       CPU LOAD
        check_command             check_load!15,10,5,!30,25,20
        use                       chapter6template
        notification_options      w,c,r
}
```

Memory

One of the things I like best about UNIX Load Average, as a metric for system performance, is that it is common across all UNIX and UNIX-like systems. Be it Linux, Solaris, or FreeBSD, the load average numbers are there and are (probably) derived in much the same way. Memory utilization,[17] however, is the opposite. Unfortunately, userspace tools, such as vmstat, don't provide an insulation layer against the differences in the various UNIX system virtual memory implementations, and other tools, such as free, might not exist on particular platforms at all. This makes it hard to write a single Nagios plug-in that accurately tracks memory across all UNIX systems.

Additionally, UNIX systems do a lot of memory caching and buffering, so even if the virtual memory systems were similar enough across the board to take these measurements, the results wouldn't be all that useful. When UNIX systems have X amount of RAM, they'll usually use most of it, and this is a good thing.

So for UNIX systems in general, the question becomes not, "How much RAM am I using?" but, "Is this system running low on RAM?" To answer that question, a much better indicator of unhealthy memory utilization is the way in which the system is using its swap space. Tools such as sar, vmstat, and iostat are very useful for querying information about swap space utilization, but again, these tools do not contain common functionality across various flavors of UNIX.

For example, vmstat on Solaris provides a page/sr column that details the page scan rate. This is a great metric for understanding whether a system is out of RAM. No matter how much memory a machine has, if the page-scanning rate is above 200 or so, you know you have a problem. Unfortunately, vmstat on Linux provides no such information, so monitoring memory in your UNIX environment depends on what kind of systems you have and what tools you have available to monitor them.

The official Nagios plug-ins tarball has a check_swap plug-in that can give you information on swap utilization for any Linux host. It is also reported to work well on BSD, but is unreliable on Solaris. Check_swap provides basic utilization info, rather than rate information, so although it can't measure things such as pages per second, it can tell you what percentage of swap is in use on a given server. To use check_swap, add the following line to your nrpe.cfg on the monitored host:

```
command[check_swap]=/usr/libexec/check_swap -w $ARG1$% -c $ARG2$%
```

This isn't much different from our previous check_load command. Check_swap will figure out the total amount of swap space available and then figure out how much is used. Thresholds are expressed as either the percentage of free swap space or the number of bytes of free swap space. The example specifies percentages by adding a percent sign after the arguments. Listing 6.28 is the command definition for the checkcommands.cfg.

Listing 6.28 *Check_swap Command Definition*

```
define command{
        command_name    check_swap
        command_line    $USER1$/check_nrpe -H $HOSTADDRESS$ \
                        -c check_swap!$ARG1$!$ARG2$
}
```

Check_swap barely scratches the surface of what you can do with memory utilization monitoring in Nagios. I highly recommend that you check the contrib. directory, as well as the Nagios Exchange, for platform-specific swap and memory utilization plug-ins.

Disk

The check_disk plug-in in the official plug-ins tarball is one of my favorite plug-ins. It's well designed and easy for lazy people to use. Check_disk can generate warnings or errors based on the free space measured in percent kilobytes or megabytes of a given device, directory, or mountpoint. Perhaps best of all, the less information you provide it, the more information it provides you. For example, here is the output of check_disk –w 10% -c 5% on one of my servers at the office:

```
DISK OK - free space: / 1725 MB (90%); /boot 76 MB (82%); /
usr 4383 MB (72%); /var 73444 MB (91%); /srv 9439 MB (94%); /
opt 1849 MB (92%); /home 9166 MB (92%);| /=198MB;1730;1826;0;1923
/boot=16MB;82;87;0;92 /usr=1664MB;5442;5744;0;6047 /
var=7191MB;72570;76602;0;80634 /srv=640MB;9071;9575;0;10079 /
opt=166MB;1813;1914;0;2015 /home=787MB;8957;9455;0;9953
```

If you don't specify any particular mountpoint directory or block device, check_disk checks all of them. This is, in my opinion, great user interface design. Performance data is provided, so a pipe splits the output with human readable output first, followed by a pipe, followed by machine parseable output. Although it's probably self-explanatory by now, Listing 6.29 is a check_disk command definition for the chkcommand.cfg file.

Listing 6.29   *Check_disk Command Definition*

```
define command{
        command_name    check_disk
        command_line    $USER1$/check_nrpe -H $HOSTADDRESS$ \
                        -c check_disk!$ARG1$!$ARG2$
        }
```

Check_MK

Check_MK, written by Mathias Kettner, belongs in both of the preceding sections as a solution both for Windows and UNIX hosts. But rather than a traditional style standalone plug-in, it operates more like a monitoring agent of the sort you might find in a large traditional corporate-style monitoring system. I know, I know—I've already done quite a bit of knocking large, traditional, corporate-style monitoring systems, but before we throw the baby out with the bath water, consider that the remote execution schemes I've covered so far have in common a substantial architectural challenge. Namely; they all serialize their checks.[18]

Because Nagios is a centralized polling system without any actual built-in monitoring logic, it must run one check at a time, on a set schedule. Nagios doesn't know if the check it is executing is being run locally or on a remote host; if it did, it might be able to parallelize its requests to remote systems—asking, for example, a remote host to execute all its service checks and return them all at once. Instead it must open 20 connections for 20 services on the same host, and quite a bit of overhead is associated with that.

Check_MK solves this problem and more, providing not just a means to parallelize all the service checks on a host, but an all-inclusive monitoring agent that dynamically detects and reports a litany of information about the host. That's right; you install the Check_MK agent on your Linux, Solaris, and Windows hosts, and without any configuration whatsoever, Nagios gets access to CPU, Memory, Disk, I/O, Network, and Thread metrics on every host.

Perhaps "without any configuration whatsoever" is strongly worded. The Nagios server must be configured with passive service check definitions for all the services Check_MK returns, but before you groan and turn the page, you should know that Check_MK creates and manages all that for you. In a way, the plug-ins dynamic configuration is the most impressive thing about it (certainly no one would use Check_MK if it wasn't there). After installing the plug-in server side and the agent on the host, you run an inventory program on your Nagios server, which dynamically detects and, through the clever use of Nagios templates, generates the complete server-side configuration for every host inventoried, including the active check for the host and the passive checks for each service detected. Although I'm thoroughly impressed by it, I don't envy Mathias writing that inventory script; that's the kind of programming very few of us enjoy. I can say from experience, however, that it works like a charm.

The agent is tiny, being a shell-script running under xinetd on Linux. Unlike NRPE, no attributes or arguments are passed from the server, which limits the vulnerability footprint. The agent is easily extended by way of a plug-in directory into which you may drop your own scripts. These custom scripts will be called by the MK agent, and assuming they follow some simple formatting rules, their output will be parsed by the server plug-in without any additional configuration. The plug-in will, in turn, generate passive checks for them and report them back to Nagios.

By default, the agent dispenses the big four food groups—CPU, RAM, disk, and network—autodetecting in the process CPU numbers, NICs (including virtual interfaces like tun/tap), and even disk partitions. A dizzying array of other data is thrown in for the bargain, including a process list and a host of hardware-specific info about devices such as Nvidia and 3-Ware cards, ACPI, and on and on. The Windows agent includes all sorts of Windows and Active Directory metrics. A full list can be had by calling the plug-in on the command line with a -M switch.

The agent program passes status to the plug-in in a way that draws a distinction between availability and performance data. The plug-in is, in turn, aware of performance data, which it can send to an RRDTool front-end for Nagios called PNP4Nagios, which I talk more about in Chapter 8, "Visualization." The plug-in even automatically generates the appropriate action_url syntax in the Nagios configuration; the graphs generated by PNP4Nagios are displayed on the Nagios web Interface, all without the Admin needing to lift a finger.

The Check_MK plug-in provides hooks to customize the configuration it generates, making it easy to specify alert thresholds for individual services on individual hosts. The rules are implemented as a cascading series of defaults, with the most specific match winning. It can also query SNMP devices like routers and switches using snmpwalk in lieu of a host-side agent.

It's possible that by parallelizing our service checks, we might learn something about how they interact. The Check_MK plug-in has a few neat features that explore this possibility, including the capability to detect the primary node in an HA-Cluster using service information returned by the agents, and a feature called Service aggregations. This latter is an attempt to capture business logic and deserves a brief explanation because it's actually quite a powerful idea.

A service aggregation can be thought of as a virtual service that is made up of several real services; for example, one can imagine a virtual service called "Email", which is made up of the qmail-send daemon on several hosts along with a few database and HTTP processes on various other hosts throughout the infrastructure. Check_MK allows you to define these virtual services and present them to Nagios as if they were real. If any of the individual services that compose them goes down, Check_MK marks the top-level virtual service down as well.

Let's take a quick look at a simple Check_MK setup. First, after downloading the current version of the Check_MK tarball from: http://mathias-kettner.de, I run the installation program like so:

```
./setup
```

After the setup program takes a look at my local Nagios installation, autodetecting the locations of the various components, it will prompt me with a few dozen questions, mostly to confirm what it already knows. After answering these questions, setup will install the various pieces of Check_MK to the appropriate locations on my Nagios server.

Next I'll install the Check_MK agent on the client system I want to monitor, which, in this case, happens to be an OpenBSD system (the installation is pretty much the same on any UNIX-style host, and on Windows it's made even easier by the Windows client installer program). To do this, I copy the appropriate agent script from the agents subdirectory of the tarball using the following command:

```
scp agents/check_mk_agent.openbsd dave@server1:
```

The agent itself is just a shell script. If I call it on the command line, I get the output of its various checks. For Check_MK to remotely execute the script, I need to run it under a tcp server like xinetd or ucspi-tcp on TCP port 6556. Instructions for setting up xinetd are available at the Check_MK website, but truthfully, I'm not a huge fan of xinetd, so the following line will run the agent under ucspi-tcp:

```
tcpserver -RXDH -l 'hostname' 0.0.0.0\  6556\/srv/mk/check_mk_agent.
➥openbsd
```

After the agent is up and running on our client, I'll tell Check_MK about it, by adding its hostname to /etc/nagios/check_mk/main.mk. Then I can tell Check_MK to "inventory" the host with this:

```
check_mk -I
```

The Check_MK script will then connect to the host and autodetect the various services available on it. I can generate the requisite Nagios configuration for these services and reload the Nagios Daemon in "one fell swoop," by calling:

```
check_mk -O
```

Or, if I want to see the config before committing it to the running Nagios daemon, I can use the "-U" switch instead. This will generate the Nagios service configuration file, usually stored in /etc/nagios/objects/check_mk_objects.cfg, but will not HUP the Nagios daemon.

From now on, the only active check I need to maintain on my client is Check_MK, and that single active check will update the status of every service running on the host that Check_MK knows about. To add services, or update the services on a running host, I call Check_MK with the –I switch to reinventory all the hosts (only new services will be detected), and either –O or –U, to commit the changes to Nagios.

Watching "Other Stuff"

Our discussion thus far has focused on traditional computer systems running some form of Windows or UNIX. In this section, I'll take a look at some of the other stuff you can monitor with Nagios, including networking gear and environmental sensors.

SNMP

SNMP was created by the IETF in 1988 as a way to remotely manage IP-based devices. It was originally described in RFC 1157 and, since then, has become as ubiquitous as it is hated by security professionals worldwide, but more on that later. The protocol is very large in scope. The IETF clearly did not intend to describe a communication protocol, but rather to provide a standardized configuration instrumentation to all IP-based devices going forward. In doing so, they not only specified strictly typed variables for configuration settings but also an all-encompassing database for network device configuration settings and status information.

Imagine, for a moment, a single hierarchal structure that gives a unique address to every piece of information and configuration setting on every IP-enabled device in the world. The hierarchy you are imagining is called the SNMP MIB tree (Managed Information Base). Each configuration setting is given an address, called an Object Identifier (OID). The OID address is similar to an IP address in that it is dot separated and gets more specific from left to right. Each number in the address specifies a node in the MIB. Let's dissect an OID example:

```
.1.3.6.1.2.1.2.2.1.8.1
```

As you can see, the OID is very much like an overly long IPv4 address. Like DNS records, the beginning "." on the left specifies the root of the tree. The first number, 1, belongs to the International Standards Organization; therefore, every node specified in this hierarchy is administered by the ISO. The 3 has been designated by the ISO for use by "other" organizations. The United States Department of Defense owns 6, and its Internet Activities Board owns 1. Every OID you'll come into contact with will probably begin with .1.3.6.1. Figure 6.1 graphically depicts this common hierarchy.

From here, the Hierarchy continues to become more specific. The three children of the IAB in Figure 6.1 are all commonly used; the children are management (2), experimental (3), and private (4). The experimental branch is used for nodes that will eventually become IETF Standards, but have not yet been. Because RFCs can take a while to become RFCs, the experimental branch enables administrators to use experimental OIDs on production gear without causing conflicts with existing OIDs. The private branch is for private organizations, such as Cisco and Juniper Networks. Private is where you would find information, such as the one-minute CPU load average on a Cisco Router. Our example OID is in the management branch, which contains IETF Standard nodes. Every node within is described somewhere in an RFC.

SNMP Hierarchy: .1.3.6.1

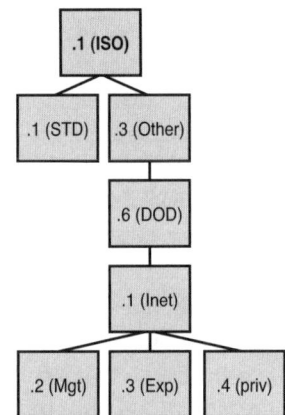

Figure 6.1   The .1.3.6.1 prefix

The next 1 places our OID within the mib-2 group, meaning that this is an SNMP Version 2 node. SNMP version 1, as first described by the IETF, had two main problems. The first was Moore's Law. When the IETF set out to specify how data could be stored in the MIB, they underestimated the size of the data values that the MIB would need to store. As a result, many of the configuration settings on the network devices quickly outgrew the IETF's strictly typed variables. The second problem was the lack of security in the SNMP protocol.

SNMP Version 1 uses a shared secret, called the community string, to authenticate peers. Unfortunately, this string is transmitted in clear text between SNMP peers, and there is a nearly universal practice of setting the string to public. This setting is often overlooked by administrators and, in years past, SNMP was enabled by default on many devices, opening them up to remote configuration by possibly malicious entities.

SNMP Version 2 was drafted in 1993 to fix these problems and add some functionality to the protocol, but unfortunately, it died an untimely death when its drafters couldn't agree on the details of the security enhancements. Three years later, a compromise was struck with the release of SNMP Version 2c. V2c took the things the IETF could agree on and combined them with the Version 1 security model. The new standard includes larger data structures, get-bulk transfers, connection-oriented traps, improved error handling, and the same flawed clear-text security model. V2c isn't as widely adopted as V1, but support for it exists on most newer network gear, including Cisco routers with IOS 11.2(6)F and later.

The IETF finished work on SNMP Version 3 in March of 2002. V3, currently a full Internet standard, is described in RFC 3410-3418. Adoption of V3 seems to be occurring more quickly than did V2c, with many vendor implementations currently available, including Cisco gear running IOS 12.0(3)T and later. V3 supports three security models: noauthnopriv, which sends clear-text SNMP packets and has trivial authentication, authnopriv, which sends clear-text SNMP packets and has strong authentication via MD5 or SHA, and authpriv, which has both encrypted sessions and strong authentication. The use of Version 3, in at least authnopriv, is highly encouraged when it's available.

The next portion of our sample OID, 2.2.1, translates to "interfaces.ifTable.ifEntry." This means our OID is in reference to a specific interface. The second-to-last digit, 8, translates to the ifOperStatus node, which is the current operational status of the interface. The very last digit in our OID specifies the interface number, so this OID translates to the current operational status of Interface #1. This value will be one of seven possible values: up, down, testing, unknown, dormant, notPresent, or lowerLayerDown.

Even though the MIB is hierarchal, devices in practice do not implement the entire tree. Devices that support SNMP contain only the subset of the MIB tree that they need. So if a given network device has no configuration settings in the experimental branch of the MIB, they do not implement that portion of the MIB tree. An SNMP device can be either a manager or an agent.[19] Agents can be thought of as SNMP servers; they are devices that implement some subset of the MIB tree and can be queried for configuration information or configured remotely via SNMP. Devices that poll or configure agents are called managers. The snmpget program from the net-snmp project is a manager, and my Cisco router is an agent.

SNMP agents don't have to sit by passively and wait to be polled; they can also notify managers of problems using an SNMP trap. SNMP traps use UDP and are targeted at the manager's port 162. Although Nagios has no intrinsic SNMP capabilities,[20] the check_snmp plug-in, combined with passive alerts and the snmptrapd daemon from the net-snmp project, make it into an SNMP manager that is capable of both polling SNMP devices for information and collecting traps from SNMP agents.

Working with SNMP

Let's get down to business and start monitoring some devices using SNMP. The first step is to get and to install the net-snmp libraries on the Nagios server. The libraries are freely available from www.net-snmp.org. After they are installed, you may have to rebuild the plug-ins to get check_snmp installed because it won't build unless the net-snmp libraries exist. When everything is installed, we can start poking around our devices with the snmpwalk program from the net-snmp package.

These days, SNMP is probably disabled on your networking gear, so before snmpwalk will be able to see anything, you have to enable SNMP somewhere. The commands in Listing 6.30 should get SNMP working in a relatively safe manner on most modern Cisco routers.

Listing 6.30   *Enabling SNMP on Cisco Routers*

```
ip access-list standard snmp-filter
permit 192.168.42.42
deny    any log
end
snmp-server community myCommunity RO snmp-filter

###########alternatively, if your router supports V3############

snmp-server view myView mib-2 include
snmp-server group ReadGroup v3 auth read myView
snmp-server user dave ReadGroup v3 auth md5 encrypti0nR0cks
```

Line 1 creates an access list called snmp-filter. The permit line specifies the Nagios server, allowing it to connect. All other hosts are denied and their attempt logged. Finally, SNMP is enabled in a read-only capacity, with the community name of myCommunity, to the hosts allowed in the access list snmp-filter. The SNMP v3 config is outside the scope of this book. Let's check out what our router has to say for itself:

```
snmpwalk -v2c -c myCommunity 192.168.42.42
```

The -v switch specifies the protocol version; -c is the community string. This command, on my Cisco 2851 router, returns 1,375 lines of output. Most of it looks like Listing 6.31, which is to say it looks like a bunch of unrecognizable SNMP gobbledygook.

Listing 6.31   *Unrecognizable SNMP Gobbledygook*

```
SNMPv2-SMI::mib-2.15.3.1.1.4.71.12.23 = IpAddress: 4.68.1.2
SNMPv2-SMI::mib-2.15.3.1.2.4.71.12.23 = INTEGER: 6
SNMPv2-SMI::mib-2.15.3.1.3.4.71.12.23 = INTEGER: 2
SNMPv2-SMI::mib-2.15.3.1.4.4.71.12.23 = INTEGER: 4
SNMPv2-SMI::mib-2.15.3.1.5.4.71.12.23 = IpAddress: 4.71.12.23
SNMPv2-SMI::mib-2.15.3.1.6.4.71.12.23 = INTEGER: 30511
SNMPv2-SMI::mib-2.15.3.1.7.4.71.12.23 = IpAddress: 4.71.12.24
SNMPv2-SMI::mib-2.15.3.1.8.4.71.12.23 = INTEGER: 179
```

```
SNMPv2-SMI::mib-2.15.3.1.9.4.71.12.23 = INTEGER: 3356
SNMPv2-SMI::mib-2.15.3.1.10.4.71.12.23 = Counter32: 4
SNMPv2-SMI::mib-2.15.3.1.11.4.71.12.23 = Counter32: 2
SNMPv2-SMI::mib-2.15.3.1.12.4.71.12.23 = Counter32: 147184
SNMPv2-SMI::mib-2.15.3.1.13.4.71.12.23 = Counter32: 147183
SNMPv2-SMI::mib-2.15.3.1.14.4.71.12.23 = Hex-STRING: 00 00
```

It looks like interesting information, if we could somehow figure out what it is in reference to. The problem is that snmpwalk can resolve only the first bit of the OIDs to their English names. This is because I lack what is referred to as an MIB file, which maps numerical OIDs to their ASCII counterparts. Net-snmp comes with MIB files for most of the management mgmt branch of the MIB tree, but you may need to download custom MIBs for OIDs in the private branch, or weirdoes in the management mgmt branch, such as the OIDs in Listing 6.30. To make sense of things, I need to install an MIB file so that snmpwalk can resolve the rest of the OID. To find the specific MIB file I require, I need to make things more numeric:

```
snmpwalk -v2c -c myCommunity -On 192.168.42.42
```

Adding the -On switch to our command causes snmpwalk to print the full OID instead of printing a partial English name followed by the part of the OID it couldn't resolve. This gets us the output in Listing 6.32.

Listing 6.32 *Even Less Recognizable SNMP Gobbledygook*

```
.1.3.6.1.2.1.15.3.1.1.4.71.12.23 = IpAddress: 4.68.1.2
.1.3.6.1.2.1.15.3.1.2.4.71.12.23 = INTEGER: 6
.1.3.6.1.2.1.15.3.1.3.4.71.12.23 = INTEGER: 2
.1.3.6.1.2.1.15.3.1.4.4.71.12.23 = INTEGER: 4
.1.3.6.1.2.1.15.3.1.5.4.71.12.23 = IpAddress: 4.71.12.23
.1.3.6.1.2.1.15.3.1.6.4.71.12.23 = INTEGER: 30511
.1.3.6.1.2.1.15.3.1.7.4.71.12.23 = IpAddress: 4.71.12.24
.1.3.6.1.2.1.15.3.1.8.4.71.12.23 = INTEGER: 179
.1.3.6.1.2.1.15.3.1.9.4.71.12.23 = INTEGER: 3356
.1.3.6.1.2.1.15.3.1.10.4.71.12.23 = Counter32: 4
.1.3.6.1.2.1.15.3.1.11.4.71.12.23 = Counter32: 2
.1.3.6.1.2.1.15.3.1.12.4.71.12.23 = Counter32: 147189
.1.3.6.1.2.1.15.3.1.13.4.71.12.23 = Counter32: 147188
.1.3.6.1.2.1.15.3.1.14.4.71.12.23 = Hex-STRING: 00 00
```

Now that we have a full OID, we can proceed to Cisco's OID Navigator at http://tools.cisco.com/Support/SNMP/do/BrowseOID.do?local=en and paste in one of the OIDs depicted in Figure 6.2.

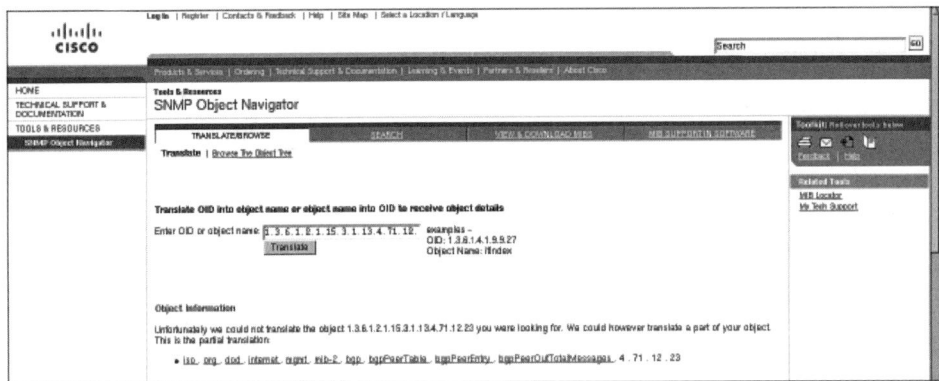

Figure 6.2   Looking up the OID at Cisco's Object Navigator web site

Now we can see the OID that we're missing: bgpPeerIdentifier. Clicking the MIB name takes you to a page where you can download the MIB as a .my file. Installing the MIB is a snap:

```
cd /usr/share/snmp/mibs²¹ && wget ftp://ftp.cisco.com/pub/mibs/v2/
BGP4-MIB.my
```

After it is installed, we can force snmpwalk to load it with the -m switch. You can give -m the name of a specific MIB to load, or specify the keyword all to load them all:

```
snmpwalk -v2c -c myCommunity -m all 192.168.42.42
```

This makes our output into something we might actually be able to use, as you can see in Listing 6.33. SNMP can be a challenge to use for the first time. There's a lot of data, and it can take some searching to find what you want if you don't know what you're looking for.²² Most administrators I know go through the process of tracking down the MIBs for their particular combination of network devices once or twice before they begin to carry their SNMP MIB collection with them from box to box, along with their vimrc, muttrc, and assorted other hard-won configuration files. Vendor-specific mailing lists are always a good source of OID tips.

Listing 6.33 *Fully MIBd snmpwalk Output*

```
BGP4-MIB::bgpPeerIdentifier.4.71.12.23 = IpAddress: 4.68.1.2
BGP4-MIB::bgpPeerState.4.71.12.23 = INTEGER: established(6)
BGP4-MIB::bgpPeerAdminStatus.4.71.12.23 = INTEGER: start(2)
```

```
BGP4-MIB::bgpPeerNegotiatedVersion.4.71.12.23 = INTEGER: 4
BGP4-MIB::bgpPeerLocalAddr.4.71.12.23 = IpAddress: 4.71.12.23
BGP4-MIB::bgpPeerLocalPort.4.71.12.23 = INTEGER: 30511
BGP4-MIB::bgpPeerRemoteAddr.4.71.12.23 = IpAddress: 4.71.12.24
BGP4-MIB::bgpPeerRemotePort.4.71.12.23 = INTEGER: 179
BGP4-MIB::bgpPeerRemoteAs.4.71.12.23 = INTEGER: 3356
BGP4-MIB::bgpPeerInUpdates.4.71.12.23 = Counter32: 4
BGP4-MIB::bgpPeerOutUpdates.4.71.12.23 = Counter32: 2
BGP4-MIB::bgpPeerInTotalMessages.4.71.12.23 = Counter32: 147205
BGP4-MIB::bgpPeerOutTotalMessages.4.71.12.23 = Counter32: 147204
BGP4-MIB::bgpPeerLastError.4.71.12.23 = Hex-STRING: 00 00
```

Now that you know what you're looking for, you can use the snmpget program from the net-snmp package to query it directly, like so:

```
snmpget -v2c -c myCommunity -m all 192.168.42.42 BGP4-MIB::bgpPeer
➥LastError.4.71.12.23
```

After you've decided on the things you want Nagios to monitor, take note of their OID numbers. I recommend that you use the numerical OID number rather than the name. This is less error prone because it removes MIB file dependencies. The check_snmp plug-in in the official plug-ins tarball is a reimplementation of the snmpget utility with some output processing functionality and Nagios return codes. Check_snmp can be a pretty complex plug-in to use, especially when text comparison becomes involved, but it packs a lot of functionality.

Let's set up a Nagios check against the SNMP attribute: iso.org.dod.internet.private. enterprises.cisco.local.lcpu.avgBusy1. This is the 1-Minute CPU Utilization Average on a Cisco PIX Firewall. Listing 6.34 is our check_snmp command definition.

Listing 6.34   *The Check_snmp Command Definition*

```
define command{
        command_name    check_fwCpu
        command_line    $USER1$/check_snmp -H $HOSTADDRESS$ \
                        -o .1.3.6.1.4.1.9.2.1.57.0 \
                        -C $USER5$ -v 2c -w 0:70 -c 0:100
        }
```

Most of this is pretty self-explanatory. The community name option, -c, from the snmpget command, became -C because check_snmp needed the lowercase version for the critical threshold. Note the appearance of the $USER5$ macro, which we haven't seen in any previous example. $USER5$ is a variable that I specified in the resources.cfg file.[23] This file allows you to define your own Nagios macros, and the nice thing about the file is its permissions. It's owned by root- and read-only, so it's a safe place to put things, such as community strings and passwords, when you need to use them in object definitions.

The warning and critical thresholds in Listing 6.34 are specified as ranges. This is something common to a few plug-ins, but this is the first time it's popped into one of the examples. The range is expressed as two numbers separated by a colon. The number on the left side is the minimum value, and the number on the right is the maximum. Every time you see this threshold syntax used by a plug-in, you may also specify the thresholds in pieces. In other words, it's possible to just specify the min *or* the max, so whereas 0:5 means "0 to 5," just :5 would mean "5 at the most, with no minimum value."

Environmental Sensors

In the past few years, environmental sensors have been popping up all over the place. With the popularity of the low cost lm78 sensor chip from National Semiconductor, just about anything with a microcontroller or microprocessor in it has some form of onboard environmental monitoring available these days. It's sometimes questionable how sensitive or how well calibrated these sensors are in low-end hardware, but most server-grade equipment possessing these sensors is usable in my experience.

The original lm78 was a special purpose microcontroller with an onboard temperature sensor and inputs for monitoring up to three fans, two voltage sources, and additional external temperature sensors. It supported two bus architectures for communication with external sensors: i2c (pronounced "i squared c") and ISA (IBM's Industry Standard Architecture). The i2c bus, created in the 1980s by Phillips, proved to be a highly successful serial bus for sensor networks because 112 i2c devices can communicate via its simple low-cost two-wire interface. Several competing architectures have sprung up since then, including Dallas Semiconductors' 1wire,[24] Motorola's SPI (Serial Peripheral Interface, pronounced "Spy"), and Intel's System Management Bus (SMBus).

Small differences exist between these bus architectures, but the choice of bus in a given system is usually a function of the brand of microcontroller and sensors involved rather than the inherent superiority of one bus over another. In practice, it's common for a single motherboard to have a combination of several buses and sensors from different vendors. Even when the motherboard has only a single sensor chip, often unexpected components contain their own. Examples include video or TV-tuner cards, battery recharging subsystems, and "backlighting" subsystems for controlling things such as LCD brightness.

If you don't trust the onboard environmental monitoring hardware, the marketplace in the past few years has exploded with standalone environmental sensors designed for data center use. It's now possible to spend anywhere from $50 for simple temperature/humidity sensors all the way up to $5,000 for camera-embedded environmental sensor arrays with features such as motion, light, smoke, and water detection, and cabinet entry alarms. If you're short on cash, several web sites sell standalone sensor kits for around $20, and a few sites will teach you how to build your own for even less.[25]

Nagios can easily interact with standalone sensors, as well as their system-embedded counterparts.

Standalone Sensors

Standalone sensors are self-contained units that are either dedicated to environmental monitoring or are part of some related piece of server-room hardware.[26] These sensors are usually very accurate and the higher-end models have some advanced functionality, such as the capability to communicate with each other to form sensor networks. Most sensors of this type, which are designed for data-center work, have Ethernet hardware and make their data available via some combination of SNMP, SSH, HTTP, and TELNET. SNMP in read-only mode is usually the preferred methodology for interacting with standalone sensors.

The first sensor that bears mentioning is, of course, the Websensor EM01B. You can't beat this sensor on Nagios compatibility; the Nagios home page has been linking to it for more than six years now. The EM01B is a standalone sensor that includes temperature, humidity, and illumination levels. It is expandable with external add-ons, such as a cabinet door alarm. The sensor has a 10/100 Ethernet card and communicates via TCP. It even comes with its own Nagios plug-ins, one written in Perl and the other in C. Expect to pay somewhere around $450.

If you need something more than a single sensor in the rack, I've heard good things about APC's EMU. The EMU is a network appliance that resembles a rack-mountable eight-port Ethernet switch. Various sensors plug into the EMUs RJ45 jacks, and multiple EMUs can be networked together to form sensor networks. Available sensors include temperature, humidity, motion, smoke, and water. The EMU can also control devices that aren't sensors, such as alarm beacons (literally, police-car-style lights meant to be mounted on the outside of the rack) and sirens. The EMU may be interfaced via a web interface, TELNET, SSH, or SNMP, so Nagios can monitor it using any of a number of plug-ins, including check_ssh and check_snmp. The EMU is a bit pricier than the EM01B.

A great place to do some research on standalone environmental sensors is Dan Klein's excellent thermd project page. Thermd is a rather complex Perl script intended to collect and plot data from various environmental sensors in your home. Dan has a lot of hands-on experience with many different power and environmental sensors, so you can be sure if thermd supports a given sensor, there's bound to be a way to integrate it with Nagios. The sensors page is at www.klein.com/thermd/#devices.

LMSensors

Using the embedded sensors inside Intel-based servers gives you a better idea of what the temperature is, where it matters the most. The lm-sensors project provides a suite of tools for detecting and interacting with all sorts of server-embedded monitoring hardware. Lm-sensors can be downloaded from the project page at www.lm-sensors.org/. While you're there, you should visit their outstanding Wiki page, which contains all sorts of great information about how to void your PC's warranty.

Users with a 2.4 series kernel may also want to install the i2c package, which is also available from the lm-sensors web site. If lm-sensors is compiled on a machine with a 2.6 series kernel, it will attempt to use the i2c support included in the kernel. Most current Linux distributions come with a copy of lm-sensors preinstalled, so you might already have it.

After lm-sensors is built, run the sensors-detect program. The sensors-detect program repeatedly warns you about the terrible things that can happen if you decide to continue and, after you repeatedly tell it to continue anyway, detects sensor chips on your motherboard and produces output suitable for appending to your modules.conf file. When you've inserted the drivers that sensors-detect says you need, you can launch the sensors program. The output of the sensors program will look something like Listing 6.35.

Listing 6.35   *Output from the Sensors Program*

```
w83627hf-isa-0290
Adapter: ISA adapter
VCore 1:    +1.52 V  (min =    +0.00 V, max =    +0.00 V)
VCore 2:    +3.36 V  (min =    +0.00 V, max =    +0.00 V)
+3.3V:      +3.41 V  (min =    +3.14 V, max =    +3.47 V)
+5V:        +5.05 V  (min =    +4.76 V, max =    +5.24 V)
+12V:      +12.28 V  (min =   +10.82 V, max =   +13.19 V)
-12V:      -11.62 V  (min =   -13.18 V, max =   -10.80 V)
-5V:        +0.23 V  (min =    -5.25 V, max =    -4.75 V)
V5SB:       +5.75 V  (min =    +4.76 V, max =    +5.24 V)
VBat:       +2.11 V  (min =    +2.40 V, max =    +3.60 V)
fan1:          0 RPM  (min =      0 RPM, div = 2)
fan2:          0 RPM  (min = 9926 RPM, div = 2)
fan3:          0 RPM  (min = 135000 RPM, div = 2)
temp1:      +34°C  (high = +40°C, hyst = +0°C) sensor=thermistor
temp2:    +25.5°C  (high = +40°C, hyst =+35°C) sensor=thermistor
temp3:    +23.5°C  (high = +40°C, hyst =+35°C) sensor=thermistor
```

```
alarms:    Chassis intrusion detection                          ALARM

beep_enable:
           Sound alarm disabled
```

The sensor chip on this server is monitoring various voltage levels, including both CPUs, the fan speed of three fans, and the temperature of three heat-sensitive resistors. The sensor chip is also tied to a cam-switch to detect when the chassis lid is removed. This is great information and is easily parsed with everyday shell tools. The official Nagios plug-ins tarball comes with a check_sensors shell script for use with lmsensors, and the contrib. directory contains a Perl script called check_lmmon for BSD users.

IPMI

IPMI (Intelligent Platform Management Interface) is an Intel specification for a hardware-based, out-of-band monitoring and management solution for Intel-based server hardware. Enterprise class servers from the big name vendors ship with integrated IPMI hardware and proprietary client software. IPMI is sometimes offered as an add-on option for commodity server hardware. The Dell OpenManage utility, for example, is a proprietary IPMI client.

IPMI operates independently of the operating system software on the system, which means that IPMI will remain available in the event of a catastrophic system failure or even while the system is powered off. As long as power is available to the system, IPMI can perform tasks, such as providing system and status information to administrators, rebooting the system, and even blinking LEDs so that remote-hands personnel can easily find the troubled system in high-density racks. IPMI implementations maintain a ring-buffer, similar to Linux's dmesg, which can provide detailed information about interesting hardware events, such as RAID-card and memory failures. IPMI implementations can even send alerts about problems to SNMP managers using SNMP traps.

Network access to the IPMI hardware is usually available either by an extra, dedicated network card specific to the IPMI hardware or by sharing a network card with the system. Client software interacts to the IPMI hardware either remotely, using the IPMI-over-LAN interface, or directly, through OS extensions, such as kernel modules.

Many open source tools exist to interact with IPMI hardware, including OpenIPMI, which is a kernel module and userspace library for local interaction. Ipmitool and ipmiutil are popular userspace IPMI query tools, which both support IPMI-over-LAN and local IPMI queries via various proprietary and open source drivers. Chris Wilson wrote a check_ipmi Nagios plug-in in Perl, which uses ipmitool. This plug-in is available from www.qwirx.com/check_ipmi.

I hope this chapter was as informative as it was fun to write. Systems monitoring is a fascinating undertaking that introduces an administrator to all kinds of cool technology that otherwise would have been overlooked in favor of the more mundane day-to-day tasks. Nagios is one of the few tools whose functionality scales linearly with its administrator's knowledge, so don't think for a moment that this chapter is an all-encompassing Nagios feature list; in fact, it is barely an overview of what the tool is capable of.

End Notes

[1]   The argument doesn't explicitly state that it is the warning threshold but, because the $ARG1$ macro is the argument to the -w switch in the check_ping command definition, the first argument listed in the service definition defines the warning threshold.

[2]   The term wrapper, in this context, refers to a program that calls another program or collection of programs to accomplish its intended purpose.

[3]   Because their security consultant told them this was the best thing to do.

[4]   I know. Where *do* I come up with these ridiculous examples?

[5]   See Tim Hill's excellent book, *Windows NT Shell Scripting*.

[6]   This is not a requirement. The PATHTEXT environment variable contains a list of extensions to which the script name is appended in the event that the extension is omitted.

[7]   The purpose it was originally designed for in 1991.

[8]   Or class path, if you prefer.

[9]   Assuming we had the privileges that we needed on the remote host.

[10] I've written them from scratch in the past, but I don't admit to being a Windows Administrator.

[11] A good friend of mine, while attending such a course, went so far as to temporarily sabotage the instructor's network connection to see what the instructor would do in the event that sample code on the Internet was unavailable. (The instructor quickly reverted to searching his local hard drive for a suitable script to modify.)

[12] Usually between 1 to 3 megabytes, depending on how many libraries the scripts include.

[13] Or something similar. I wasn't able to find any information on the actual data structure behind import-csv.

[14] As opposed to what, I wonder? Expert XML manipulators? DCOM OOP specialists?

[15] Believe me, I've tried.

16 The fact that the feature is doubly disabled by default and called "Don't blame us" in the config file should give you pause. Use it at your own risk and don't blame me, either.

17 Check out the Solaris section of the Nagios Exchange (www.nagiosexchange.org) for Solaris-specific plug-ins.

18 This isn't entirely true. The developer of NSClient++ is currently working on a special-purpose protocol that will allow service-check parallelization.

19 Or, in some cases, both.

20 Nagios has few intrinsic capabilities of any type and that is a good thing. See Chapter 2.

21 This path is almost certainly something different for you. Check your net-snmp installation packaging.

22 If you have a load balancer, check out http://vegan.net/MRTG/index.php; it's a sanity-saving collection of often-used SNMP metrics for all sorts of load balancers.

23 As described in Chapter 4, "Configuring Nagios."

24 Which, because of the need for a ground wire, actually uses two wires

25 For the ultimate in DIY hardware-based server monitoring, check out Bob Drzyzgula's work: www.usenix.org/publications/library/proceedings/lisa2000/drzyzgula.html.

26 I'm personally enamored with a cool power strip in one of my company's collocation facilities. The PowerTower XL has 32 programmable outlets on two internally redundant 20-amp busses and two external temperature/humidity probes on six-foot wires. I especially enjoy telling my friends about how I periodically SSH into my power strip.

## CHAPTER 7

Scaling Nagios

The load on a well-built monitoring system will inevitably outgrow its hardware. Good monitoring systems are too useful to ignore, and service by service, host by host, you will eventually find yourself in the unenviable position of having to migrate everything to a bigger, or at least beefier, server. But before you start backing up your Nagios config files, consider taking a more future-proof route. In this chapter, I explore a few options for scaling your existing Nagios install using distributed architectures—that is, spreading the monitoring load across more than one Nagios server.

Tuning, Optimization, and Some Building Blocks

Before we get into the distributed solutions, I should mention that there's quite a bit of systemic tuning you can do to lighten the load on a running Nagios daemon. These tuning parameters change between versions, so rather than redocument them all here, I'll point you to the official tuning documentation:

```
http://nagios.sourceforge.net/docs/3_0/tuning.html
```

If you're running a Nagios system that's writing data to RRDTool databases, as described in Chapter 8, "Visualization," you would do well to use an RRDTool glue layer that supports rrdcached, which is a daemon included with RRDTool that receives updates to existing RRD files, accumulates them in RAM, and when enough have been received or a defined time has passed, writes the updates to the RRD file all at once. Performing RRDTool updates in bulk instead of writing the results of every check result to disk immediately can be a huge performance win.

NRDP/NSCA

As described in Chapter 2, "Theory of Operations," Nagios services can be configured as passive by setting the `active_checks_enabled` attribute to 0, and `passive_checks_enabled` to 1. After this is done, Nagios will not schedule any checks for the passive service, expecting instead to receive passive check results from some external monitoring system by way of the command FIFO. If you have tertiary monitoring systems already in place you can use them to update services in Nagios, thereby saving Nagios the overhead of scheduling and executing the checks.

Instead of adapting external monitoring systems to use the Nagios FIFO directly, most administrators rely on either NSCA (the Nagios Service Check Acceptor) or NRDP (Nagios Remote Data Processor) to submit passive check results. Both of these tools come with a server daemon that listens on the Nagios server and client programs that run on remote hosts. The client is used to push passive check results from remote hosts, such as other monitoring systems, to Nagios, where they are injected into Nagios's reaper queue.

NRDP is the newer of the two tools and is generally considered to be NSCA's replacement. It runs as a PHP web application on the Nagios server and can therefore be reached on the usual HTTP ports. Rather than injecting commands into the command FIFO, it bypasses the FIFO and uses an event broker module to inject commands directly into the Nagios reaper queue. The NRDP clients have been written in Perl, Python, PHP, and bash. Service check results can even be submitted manually by directing a browser at the NRDP PHP program.

NDOUtils

If your Nagios system is struggling under the load of a homegrown data export solution built from scripts that, for example, parse the status.dat, or use event handlers to send data to external systems, you might consider replacing it with NDOUtils. This core plug-in exports the state of a running Nagios daemon to a MySQL database, which can be run on a remote system. Although some overhead is associated with this, it's usually cheaper than solutions based on parsing the status log or running shell scripts from event handlers.

Distributed Passive Checks with Secondary Nagios Daemons

The first methodology I'm going to cover makes use of the "passive checks" feature I've described previously. As depicted in Figure 7.1, this arrangement allows us to add a Nagios server to our environment configured so that instead of processing its own check results, it submits them as passive check results to another, parent Nagios server. I'm including this method in the interest of being thorough, because although this method was once a popular means of creating distributed architectures, there are now far better ways to achieve the same result, as you'll read next.

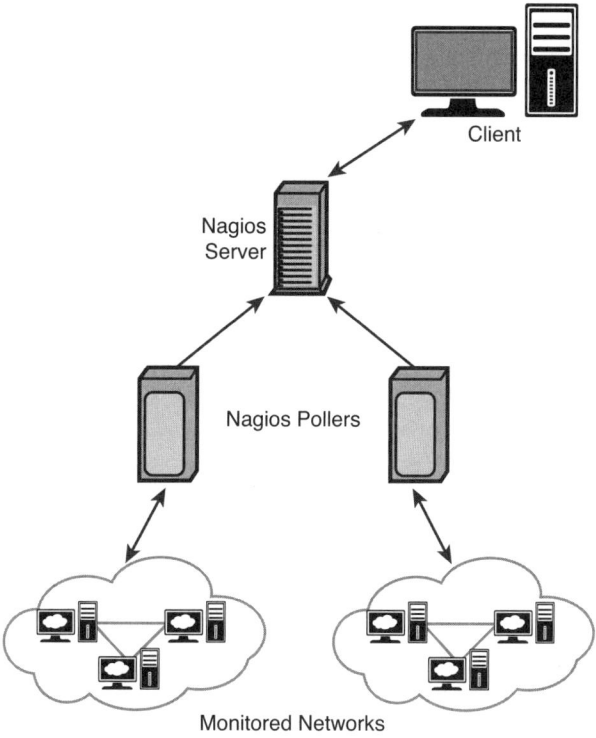

Figure 7.1   Distributed architecture using passive checks

In this methodology, the parent Nagios server is configured normally with the exception that active checks are disabled for the services that are being run by the child or "poller" Nagios instance. Let's take a look at the service configuration differences for the check_ping service under this arrangement. First, the parent configuration (see Listing 7.1):

Listing 7.1   *Ping Service Definition for the Parent Instance*

```
# Service definition for ping on child
define service{
        use                             generic-service ;
        host_name                       host1 ;
        service_description             PING  ;
        contact_groups                  admins  ;
        notification_options            w,u,c,r  ;
```

```
        notification_period              never ;
        check_command                    check_ping   ;
        active_checks_enabled               0        ;
        passive_checks_enabled              1        ;
        notifications_enabled               1        ;
}
```

Next, the child configuration (see Listing 7.2).

Listing 7.2   *Ping Service Definition for the Child (Poller) Instance*

```
# Service definition for ping on child
define service{
        use                      generic-service ;
        host_name                host1 ;
        service_description      PING  ;
        contact_groups           admins   ;
        notification_options     w,u,c,r  ;
        notification_period      never ;
        check_command            check_ping   ;
        active_checks_enabled        1        ;
        passive_checks_enabled       1        ;
        notifications_enabled        0        ;
}
```

Here we see the first drawback of using passive service checks to create distributed architectures—namely, that the service configuration must exist on both the child and parent, making this model tedious and unwieldy to configure and maintain. After the service is configured on both monitoring systems, we configure the child to send a passive check result on every service back to the parent using the Obsessive Compulsive Service Processor feature in the nagios.cfg (see Listing 7.3).

Listing 7.3   *OCSP Configuration in the nagios.cfg on the Child*

```
obsess_over_services=1
ocsp_command=submit_service_check_to_parent
ocsp_timeout=5
obsess_over_hosts=1
ochp_command=submit_host_check
ochp_timeout=5
```

Then we must define the `submit_service_check_to_parent` command in the misccommands.cfg file. This command will use the NSCA client to send the passive check result to the parent, assuming that the parent host is running the NSCA daemon (see Listing 7.4).

Listing 7.4 *The submit_service_check_to_parent Definition on the Child*

```
define command
{
command_name    submit_service_check
command_line    /usr/nagios/libexec/submit_service_check.sh \
$HOSTNAME$ '$SERVICEDESC$' $SERVICESTATEID$ '$SERVICEOUTPUT$'
```

Finally, the shell script referenced in the description is a one-liner that looks like this (see Listing 7.5).

Listing 7.5 *The submit_service_check.sh Shell Script on the Child*

```
#!/bin/bash
/usr/bin/printf "%s\t%s\t%s\t%s" "$1" "$2" "$3" "$4" | /usr/lib/
nagios/plugins/send_nsca -H  -c /usr/lib/nagios/send_nsca.cfg
```

As previously noted, the NSCA protocol is being slowly phased out in favor of NRDP, which is a more flexible alternative that uses HTTP and bypasses the Nagios command file. Not being a fan of passive service checks myself, I don't cover NSCA or NRDP very extensively in this book, but a lot of info about both protocols is available from the official add-ons page at:

```
http://www.nagios.org/download/addons/
```

Aside from the redundant configuration problems mentioned previously, the passive check solution is problematic in the event of a node failure of either the parent or the child. If the child server goes down, outages in the section of the network monitored by that child will go un-noticed, and if the parent server goes down, the children are not of much use. Further, the scalability of the entire solution (assuming NSCA is employed instead of NRDP) depends on the I/O capability of a single 64KB FIFO on the parent server, because all passive check results must traverse the command FIFO. Large installations will eventually be burdened by this limitation.

Event Broker Modules: DNX, Merlin, and Mod Gearman

The second methodology for creating distributed architectures between Nagios servers is the use of event broker modules to enable Nagios daemons running on different machines to cooperate by exchanging information with each other using protocols designed for that purpose. DNX (Distributed Nagios Executor), Mod Gearman, and Op5 Merlin are excellent examples of this design.

DNX is lightweight and simple, enabling the creation of distributed architectures that scale far beyond what's possible with passive checks. Further, DNX requires very little configuration. In fact, you only need to run Nagios on a single host; the other hosts need only a copy of the plug-ins.

Mod Gearman is similar in design to DNX, but adds customizable worker queues and a super-efficient distributed application framework. It is more to configure, but it allows you to create special-purpose worker nodes to suit your organizational needs.

Op5 Merlin is a far more complex tool with a loftier goal. In the words of its authors, Merlin is a "cross-host event processing system." It's capable of creating intricate parent-child and peer relationships between running Nagios daemons and exporting the state of a live Nagios server to a mySQL database.

DNX

We begin our discussion with DNX, which intercepts and distributes Nagios service checks to a cluster of "worker nodes." The primary node is the only host that needs to run Nagios. It loads the DNX event broker module, which in turn starts three services—the dispatcher, the collector, and the timer—and then lies in wait for the Nagios daemon to execute a service check.

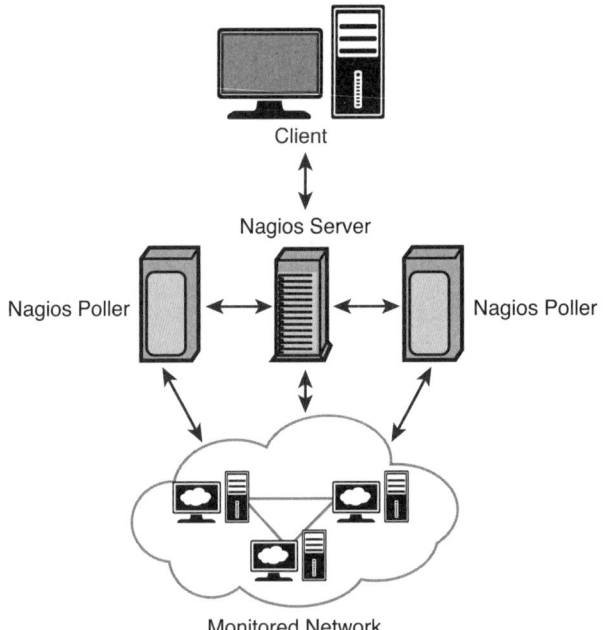

Figure 7.2   Distributed service checks with DNX

Each worker node runs a DNX daemon, which periodically sends a request for a job to the dispatcher service on the Nagios server—the "master" node. The dispatcher service on the master node keeps a list of job requests from the workers. If the Nagios daemon executes a service check, and there are no job requests available from the workers, DNX returns a 0 result code, and Nagios executes the plug-in as it normally would.

New worker nodes can be brought online at any time, and can begin working immediately without affecting the execution of the master node; no daemon restart or configuration change required on the master. The job requests sent by the worker nodes eventually time out, so in the unlikely event that every worker node dies, the Nagios server will seamlessly take over execution of the service checks as if the DNX module had never been loaded.

Assuming job requests are available, the dispatcher service intercepts the service checks from Nagios and sends them off to worker nodes, which execute the check and return the results back to the master servers DNX collector service. If the worker node dies, so do its work requests, so new checks won't be sent to it, and checks that it may have been executing when it died will be rescheduled by Nagios when no results are returned within the check timeout. The collector service on the master node bypasses the Nagios command file, injecting its results directly into Nagios's results ring buffer data structure. It is therefore not limited by I/O to the command FIFO like the passive check methodology is.

It probably took you longer to read that description than it will to get DNX up and running. Simply download the tarball and build the server on your Nagios server, the client on your worker nodes, and add the following line to your nagios.cfg file:

```
broker_module=/path/to/nagios/lib/dnxPlugin.so \ /path/to/nagios/etc/
dnxServer.cfg
```

Finally, edit the dnxClient.cfg file on the workers to specify the server host address, and start the worker process with the included init script `dnxcld`.

DNX ships with a command-line management app called `dnxstats`, which allows you to perform management tasks on workers as well as the server itself. You can use it to shut down or reconfigure nodes, as well as put them into and take them out of debug mode. As its name suggests, you can also glean various stats from the workers and server, many of which are interesting and useful and would make good fodder for time-series graphs.

Mod Gearman

Mod Gearman, an event broker module for distributing the monitoring workload of a Nagios server, gets its name from an underlying library called Gearman. Gearman is a generic application framework that is designed to help developers easily create large distributed parallel applications.

Architecturally, Mod Gearman is very similar to DNX. In fact, if I were to draw an architecture diagram of Mod Gearman, it would look identical to the one I've already drawn of DNX, so I'll refer you back to Figure 7.2 for a description of the architecture. Both solutions are event broker modules that intercept host and service checks, sending them to an external daemon process that in turn interacts with one or more worker nodes for the purpose of distributing the host and service check load.

Functionally, however, Mod Gearman is far more configurable than DNX, which can make it more difficult to deploy. Gearman introduces the concept of user-configurable queues into which service and host checks are placed. These queues can be named after Nagios servicegroups or hostgroups, and individual worker nodes can be assigned to specific queues. This enables a high-degree of job control, allowing you, for example, to dedicate a few powerful worker nodes to the most computationally demanding service checks, or specify regional worker nodes to perform checks on the hosts in their closest proximity.

Additionally, Mod Gearman has a few features DNX does not, such as an embedded Perl interpreter and tools such as send_gearman and send_multi, which allow you to implement high-performance passive checks replacing NSCA and NRDP. Mod Gearman can also distribute event handlers in addition to service and host checks.

Perhaps the most intriguing aspect of Mod Gearman, and certainly the one that makes it so popular, is its efficiency. Mod Gearman is a better tool for running host and service checks than Nagios, to the extent that you can increase the performance of a running Nagios system by installing Mod Gearman and a single worker on the Nagios server itself.

To get it running, download and untar the latest version from consol labs (labs.consol.de). Run the typical, configure, make, and make install from the source directory, and add the broker line to your nagios.cfg:

```
broker_module=.../mod_gearman.o server=localhost:4730
eventhandler=yes services=yes hosts=yes
```

The server option refers to the network location of the external daemon process "gearmand." This daemon is actually provided by Gearman framework, which must be installed before Mod Gearman will compile. Gearmand sits between the Mod Gearman NEB module and the worker clients, passing work to the clients and results back to Nagios via the event broker module. The clients should be started before the server, like so:

```
./mod_gearman_worker --server=localhost:4730 --services -hosts
```

The server may now be started with this:

```
/usr/sbin/gearmand --threads=10 --job-retries=0
```

Although this simple configuration should be enough to get you going, a vast array of options can be passed to the broker, server, and worker processes, including a "config=" option, which will cause the various processes to read their configuration from a file rather than via the command line. For full documentation on the configuration, see http://labs. consol.de/lang/de/nagios/mod-gearman/.

Op5 Merlin

The Nagios Event Broker allows third-party developers to hook into pretty much any aspect of a running Nagios daemon by registering for "callbacks"; events that are triggered when the Nagios daemon performs some action, such as sending a notification. The expectation is that the event broker module written by the third-party developers will register for the types of events it is interested in, and then when those events take place, the module will do something interesting with them. DNX, for example, registers for the NEBCALLBACK_ SERVICE_CHECK_DATA callback and uses the NEBTYPE_SERVICECHECK_INITIATE event to preempt service check execution and insert its own load-balancing framework.

However, rather than registering for a particular callback and writing handler functions to do something interesting with them, the Merlin NEB module simply registers for ALL callbacks and exports the resulting events it receives from Nagios to an external daemon. The Merlin daemon gets events from the Merlin module running inside Nagios and either sends them to other Merlin daemons on other systems or to a database of your choosing.

Events that come from other Merlin daegmons can be injected back into a running Nagios daemon via the Merlin Event Broker Module. So Merlin makes two very powerful things possible. The first is database synchronization, which in turn enables all manner of third-party UIs, add-ons, data export, and backup scenarios. The second, and more topical for us, is load balancing, clustering, and failover.

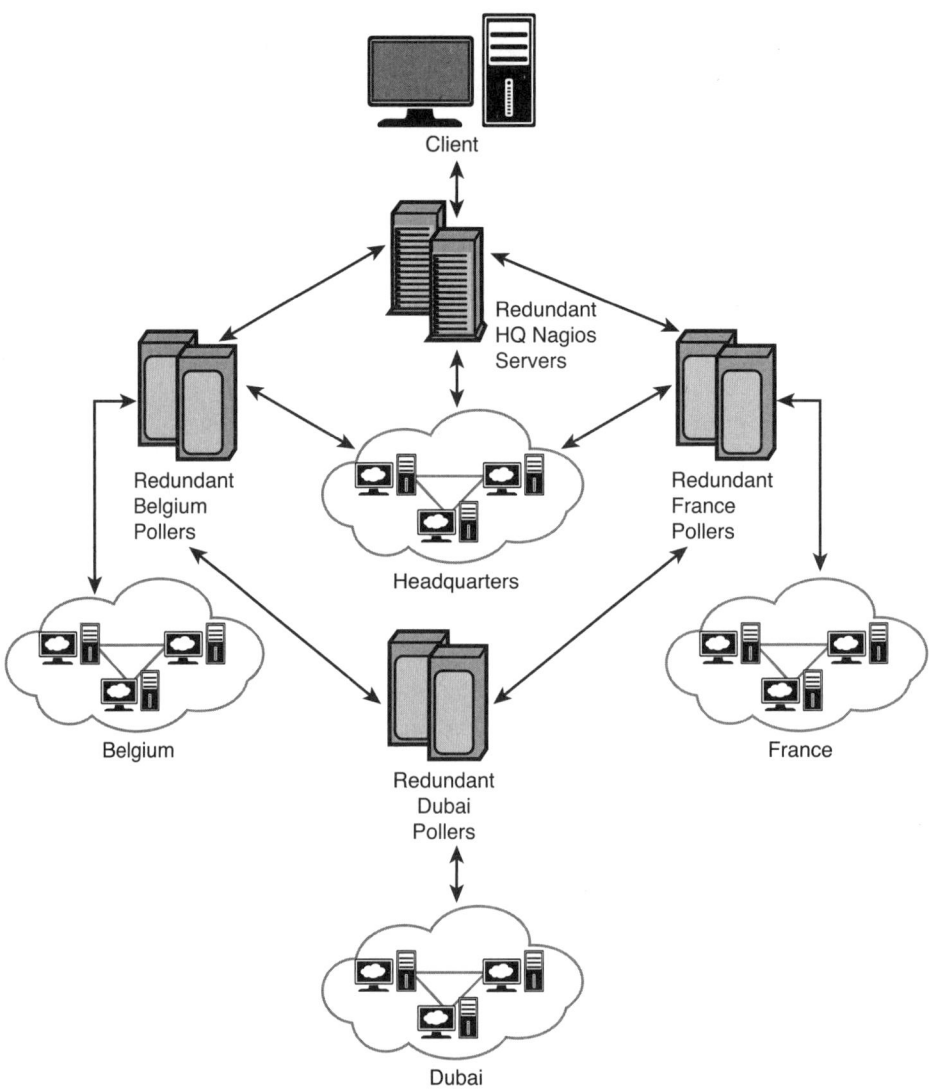

Figure 7.3   Distributed state replication with Merlin

With Merlin, it's possible to update the state of one Nagios daemon with events generated by another, remote Nagios daemon. The possibilities here are pretty interesting. DNX-like arrangements may be created where "master" Nagios daemons send their checks to subservient Nagios daemons for processing, peering clusters may be set up where two or more Nagios daemons cooperate and update each other, or various permutations of the two may be achieved. A single Nagios daemon, for example, may be a peer to its peers, a master to its nodes, and a node to its masters, all at the same time. Display-only relationships are also possible whereby, for example, a server in Chicago might collect and display the state of remotely administered daemons in India, Brazil, and Australia.

Merlin is part of a commercial product offered by Op5 called "Monitor," and although it is open-source and may be downloaded and used for free, it is not for the faint of heart. I've personally done the install four times, and every time I've spent hours wrestling with broken assumptions in the install script in the GIT repository. After you finally get it installed, it is configured via the /opt/monitor/op5/merlin/merlin.conf file. A load-balanced peer relationship between server1 and server2 looks like this:

A Merlin Load-Balanced Peer Configuration

```
#on server 1
peer server2 {
        address = server2.my.com
        port = 15551
}

#on server 2
peer server1 {
        address = server1.my.com
        port = 15551
}
```

This looks pretty simple, but things quickly become more complicated as you add dedicated pollers and so on. The Merlin configuration is slightly different on all nodes; however, every node needs to have the same Nagios configuration. If you purchase the commercial product, you can use a tool called "mon," which can generate proper configuration, distribute it to all Merlin nodes in the cluster, and synch the Nagios configuration from a single central point, which vastly simplifies Merlin's setup. With the free version, you're stuck managing the configuration across all the machines manually, which also means keeping track of a lot of minutiae, such as having to start service "x" on box "y" before starting service "b" on box "a." If you're considering Merlin for a corporate production environment and you want to avoid a splitting headache, you should seriously consider purchasing Op5 Monitor.

Distributed Dashboards: Fusion, MNTOS, and MK-Multisite

The third and final set of tools employ a strategy that can't really be called "distributed" because they do not provide a means for distributing load to multiple Nagios servers. Instead, they merely provide a common interface to several independent Nagios servers. As depicted in Figure 7.4, the Nagios servers are each configured to monitor their portion of the network, and are completely unaware of each other's existence. Integration between the various Nagios systems is provided by a web front-end that presents a summary of their collective status.

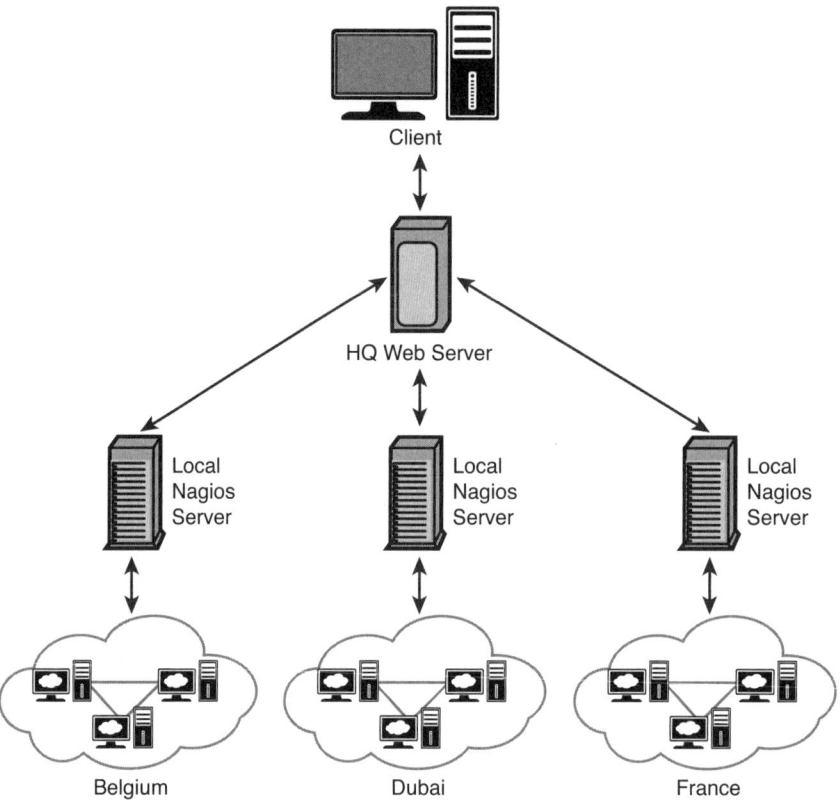

Figure 7.4 Centralized presentation of independent, distributed monitoring systems

The first two tools, Fusion and Multi Nagios Tactical Overview System (MNTOS), scrape information off the Nagios web-UI from the various servers via HTTP or HTTPS in much the same way a user would. The primary difference between them is that Fusion is a commercial product and was designed to work with Nagios XI. MNTOS, on the other hand, is open-source and is intended for use with the Nagios core. If you run Nagios XI, fusion provides an integrated experience. It has an XI look and feel; from the fusion UI, you can see the status of your entire monitoring infrastructure and click through to individual problems on individual hosts without having to relog in to the XI web-UIs. MNTOS is not much more than its name suggests. It is a replacement tactical overview that is capable of displaying the status of more than one Nagios server.

MK-Multisite, written by Mathias Kettner, is a different and far more interesting beast. Instead of scraping the Nagios web-UI, multisite depends on an event-broker module, also written by Mathias Kettner, called "Livestatus." Livestatus may be the most revolutionary event broker module written to date. Simply stated, Livestatus provides you a UNIX socket (not a network socket) through which you can query state information directly from a running Nagios daemon. You ask questions via a simple query language and receive immediate answers directly from the same data structures resident in memory that are in use by the Nagios daemon. No overhead is associated with replicating state or querying external databases. There is no Disk I/O whatsoever. There is no latency and no risk of the data being out of date, and very important, there is no risk of blocking the daemon process with a hung ODBC thread or something similar.

Livestatus is included in the check_mk tarball, which I covered in Chapter 6, "Watching: Monitoring Through the Nagios Plug-ins," but it can also be downloaded as a standalone package. If you're installing check_mk, answer "yes" when the setup script asks if you'd like Livestatus installed. The standalone build is the autoconf standard: configure, make, and make install. Either way, Livestatus.o will be compiled and placed in /usr/lib/check_mk by default. Edit the nagios.cfg to enable the event broker with the following:

```
event_broker_options=-1
```

and tell Nagios to load Livestatus with this:

```
broker_module=/usr/local/lib/mk-livestatus/livestatus.o /var/lib/
nagios/rw/live
```

Then restart Nagios and you're all set. The second argument is the location of the Livestatus socket on the local filesystem. Again this is a Unix IPC socket, so you'll need a tool like socat or unixcat to work with it from the shell. Livestatus comes with its own copy of unixcat, which it will install for you in /usr/local/bin by default. I'm unsure whether this is the same tool as the one distributed in ucspi-unix, or if it's original code. Caveat emptor.

The suggested method for making remote queries to livestatus is by piping them over SSH, but there's no reason why the UNIX socket couldn't be front-ended by a TCP server program such as xinetd or DJB's tcpserver. Sample configuration for xinetd is available at http://mathias-kettner.de/checkmk_livestatus.html.

Several other options may be passed to the NEB module on the broker_module line to, for example, place it in debug mode or tune the various cache, buffer, and thread sizes. These are fully documented at the URL I mentioned previously. After Nagios is restarted, check to make sure the socket exists in the expected location, and test that it's working with the following:

```
echo 'GET columns' | /usr/local/bin/unixcat /var/lib/nagios/rw/live
```

If everything has gone according to plan, a dizzying block of text should scroll by. The "GET" statement in the preceding command was an "LQL" or Livestatus Query Language command. LQL commands are made up of two parts; the first line, which always begins with "GET", identifies the object type you're interested in inspecting. The Livestatus documentation refers to these as "tables," but they roughly correspond to object types in Nagios. The available tables follow, shamelessly pasted from the official documentation:

- **hosts**—Your Nagios hosts.
- **services**—Your Nagios services, joined with all data from hosts.
- **hostgroups**—Your Nagios host groups.
- **servicegroups**—Your Nagios service groups.
- **contactgroups**—Your Nagios contact groups.
- **servicesbygroup**—All services grouped by service groups.
- **servicesbyhostgroup**—All services grouped by host groups.
- **hostsbygroup**—All hosts group by host groups.
- **contacts**—Your Nagios contacts.
- **commands**—Your defined Nagios commands.
- **timeperiods**—Time period definitions (currently only name and alias).
- **downtimes**—All scheduled host and service downtimes, joined with data from hosts and services.
- **comments**—All host and service comments.
- **log**—A transparent access to the Nagios logfile (including archived ones).
- **status**—General performance and status information. This table contains exactly one dataset.
- **columns**—A complete list of all tables and columns available via Livestatus, including descriptions.

As you can see, these are mostly Nagios object types. You may have noticed "logs" in the list. Your eyes do not deceive you; Livestatus can retrieve log messages for you, even if they've been archived.

The second part of an LQL command is made up of any number of subsequent lines, collectively referred to as "headers." Headers generally allow you to modify your query, restricting the output to a given set of columns, filtering it for objects that meet a criteria, transforming the output into a statistic (for example, returning how many objects matched, or computing the average of a returned value from several hosts), and even modifying the output format (to JSON or CSV for example).

I used 'GET columns' as the initial test command because a command like 'GET hosts' will return a massive amount of data from a moderately sized Nagios server. Now with the help of the "Columns" header, we can return the host attributes we're interested in, like so:

```
Q='GET hosts
Columns: address state
' echo "${Q}}" | unixcat /var/lib/nagios/rw/live
```

The "Columns" header causes Livestatus to return only the attributes you specify. The preceding query should return a list of the address of every host monitored by Nagios, followed by its current state. The returned list will use semicolons for line separators. Because LQL queries are themselves multiline, things get a little weird from the command line. In the previous example, I'm setting a variable to the multiline query and then echoing it to unixcat (the double quotes around the variable in the echo statement are important. Without them, the shell will eat the line-feeds in the variable).

The official documentation recommends that you save the query to a file, and then redirect it to unixcat like so:

```
echo 'GET hosts
Columns: address state' > query.lql

unixcat /var/lib/nagios/rw/live < query.lql
```

I quickly tired of both of these methods and took a minute to write a small shell wrapper that I call "qls" (query Livestatus) to make it easy to interact with Livestatus. It works kind of like an interactive Livestatus shell, and is shown next (see Listing 7.6).

Listing 7.6   *My qls Script, an Interactive Shell for MK-Livestatus*

```
#!/bin/sh

UC='/usr/local/bin/unixcat'
LSS='/var/lib/nagios/rw/live'
```

```
function appendHeader {
echo 'ah called'
H="${H}
${IN}"
}

function send {
echo "${Q}
${H}" | ${UC} ${LSS} | less
}

function usage {
Q="Usage:
Anything that begins with GET will be interpreted as line 1
Anything containing a ':' will be interpreted as a header
Add as many headers as you want, clear them with 'hd'
once the query looks good, type '.' to send it to livestatus
'usage' will bring back this blurb (but will also clobber your line
1)
ctl-c or 'quit' to quit
cheers"
}

unset H
usage

while [ 1 ]
do
     clear
     echo "${Q}"
     echo "${H}"
     echo '-----------'
     echo -n ": "

     read IN

     case $IN in
     GET*)
       Q="${IN}"
      ;;
     [hH][dD])
       unset H
      ;;
     *:*)
       [ "${H}" ] && appendHeader || H="${IN}"
      ;;
     .)
          send
     ;;
     usage)
       usage
      ;;
     quit)
```

```
    exit 0
    ;;
    *)
        clear
        echo "WTF?!"
        sleep 1
    ;;

    esac

done
```

If I wanted to narrow my list to just the hosts that were in a critical state, I could modify my query accordingly:

```
GET hosts
Columns: address state
Filter: state = 2
```

I can continue to narrow and modify my results with as many headers as I'd like, placing each new header on a subsequent line. The "Filter" header is quite powerful, supporting 12 equality operators, including negation, and case-sensitive and not-case-sensitive POSIX extended regular expressions ("~" for case sensitive, and "~~" for not case sensitive). I could, for example, refine my search to all the hosts whose description began with "DB" followed by a number between 1 and 4 (inclusive) with:

```
GET hosts
Columns: address description state
Filter: state = 2
Filter: description ~ ^DB[1-4]
```

Filter headers can be combined by suffixing them with a logical AND or OR header. This makes it possible, for example, to return the DB hosts that are in either a critical or warning state, like so:

```
GET hosts
Columns: address description state
Filter: state = 1
Filter: state = 2
OR : 2
Filter: description ~ ^DB[1-4]
```

The ordinal after the OR header defines how many of the previous headers are included in the logical OR.

Sometimes you don't want a list of the actual things, but rather the number of things that meet your criteria. The "Stats" header can not only provide counts, but also compute sums, averages, and standard deviation, as well as return Min and Max, and much more. The following query returns the number of hosts in a critical state:

```
GET hosts
Stats: state=2
```

Stats headers also support logical operators. Here is the number of hosts that are in a critical state whose state has not been acknowledged:

```
GET hosts
Stats: state=2
Stats: acknowledged=0
StatsAnd: 2
```

The last example I'll spare space for is an example of the "Stats" headers grouping support. By adding a "Columns" header to my Stats query, I tell Livestatus to output a line each time the given Column value is different. For example, the following query will break down the number of unacknowledged critical hosts per hostgroup:

```
GET hosts
Stats: state=2
Stats: acknowledged=0
StatsAnd: 2
Columns: groups
```

Livestatus is a pretty amazing piece of software that is worth installing on your Nagios systems, even if you don't plan to use Multisite. It transforms Nagios Core into an introspective entity, imbuing it, with a degree of intelligence it did not have before, and adding to it a powerful scripting interface. In the context of Multisite, Livestatus is not only a far more elegant solution than Fusion and MNTOS for creating a dashboard display of multiple Nagios servers, it is also far more functional, because Livestatus gives Multisite the capability to send commands to the Nagios daemons that it monitors. This makes Multisite capable of doing things like scheduling downtime for entire service groups that span multiple Nagios servers across the enterprise. This feature is something that's available on no other Nagios UI (including XI). Multisite is also capable of displaying PNP graphs and far more host and service meta-information than is available in even the native Nagios UIs.

## CHAPTER 8

Visualization

If you wanted to summarize the intent of systems monitoring in two words, it would be difficult to do better than "increase visibility." Until now we've focused on our ability to detect problems with the machines that make up our network, but a good monitoring system shouldn't stop there. Good monitoring systems act like transducers in electronics, converting the incomprehensibly large number of interactions between systems and networks into an environmental compendium fit for human consumption. They provide an organic interface to the network, which allows us to better understand its inner workings, and the extent to which they improve our visibility determines their usefulness as tools.

Data visualization is key to improving our visibility because, if done correctly, it effectively communicates the status of the environment, enables pattern recognition in historical and real-time data, and transforms propeller-head metrics into indicators laymen can utilize. It can aid any number of critical undertakings like capacity planning, root cause analysis, and even marketeering. Data visualization can catch problems you didn't tell the monitoring system to look for. Also, everybody loves a pretty picture.

Out of the box, Nagios doesn't draw many pretty pictures. The web interface has good functional data visualization, but it is focused on the current environmental state and lacks historical time series data on specific services and hosts. However, in the same way its lack of built-in monitoring logic enhances its flexibility, the lack of integrated data visualization has allowed Nagios to thrive where other systems cannot. By focusing on making the data available to external programs, Nagios arms us with what we need to use the very best data visualization software available, rather than forcing us to settle for mediocre built-in functionality. In this chapter, I'll help you integrate Nagios with various popular visualization packages.

Nagios Performance Data

Nagios plug-ins, as covered in Chapter 2, "Theory of Operations," return one of four states: 0 for OK, 1 for Warning, 2 for Critical, and 3 for Unknown. The Nagios daemon reacts to these return codes, notifying administrators via email or SMS, for example. In addition to the codes, the plug-ins may also return a line of text, which will be captured by the daemon, written to a log, and displayed in the UI. If the daemon finds a pipe character in the text returned by a plug-in, the first part is treated normally, and the second part is treated as performance data.

Performance data doesn't really mean anything to Nagios; it won't, for example, enforce any rules on it or interpret it in any way. The text after the pipe might be a chili recipe for all Nagios knows. The important point is that Nagios can be configured to handle the post-pipe text differently than pre-pipe text, thereby providing a hook from which to obtain metrics from the monitored hosts and pass those metrics to external systems, without affecting the human-readable summary provided by the pre-pipe text.

Nagios's performance data handling feature is an important hook. Quite a few visualization systems depend on it to export metrics from Nagios. These systems typically point the service_perfdata_command attribute in the nagios.cfg at a script that will use a series of regular expressions to parse out the metrics and metric names, and then store them in a database of some kind. To use these systems, you must enable performance data processing in Nagios by setting process_performance_data= 1 in the nagios.cfg file.

RRDTool: The Foundation

The overwhelming majority of the visualization systems I'm describing use a tool called RRDTool to store Nagios performance data. RRDTool makes it easy to create time-series graphs of previously stored metrics. These graphs are useful for all sorts of things because they allow us to spot trends over time and tell us what a given service was doing at a particular time in the past. Take the CPU load graph in Figure 8.1, for example.

We can clearly see a CPU spike across all servers at around midnight in the graph. If we change the time interval on the graph so that it displays a longer period of time, such as Figure 8.2, we can see that this behavior is not typical for these servers; therefore, we may surmise that this behavior may be indicative of a problem.

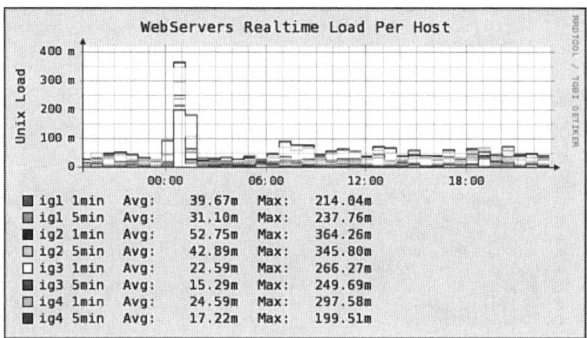

Figure 8.1   CPU load graph, past 28 hours

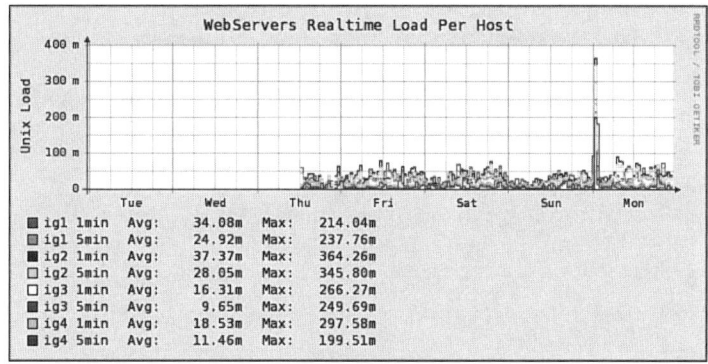

Figure 8.2   CPU load graph, last week

Time-series graphs are great for visually stimulated humans; they depict lots of information in a small space and in a way that communicates the subject matter instantly. The simple act of graphing every metric you can helps out in all sorts of situations. You can never graph too much. For instance, a comparison of Figure 8.1 to the graph of network utilization on the same servers, as shown in Figure 8.3, shows that the CPU spike coincided with a 6MB/s network spike, implying that the abnormal utilization was somehow related to traffic received by the hosts in question.

This time-series data is the first, and most important, data visualization you should add to your Nagios web Interface. There are three pieces to this puzzle: First, you need to collect the data, then store it, and finally, display it. There are dozens of visualization systems out there to help you do these things. Some of them take care of only one piece of the puzzle, and others try to do it all for you. I tend to choose from the former class of tools, unless there is a compelling reason not to.

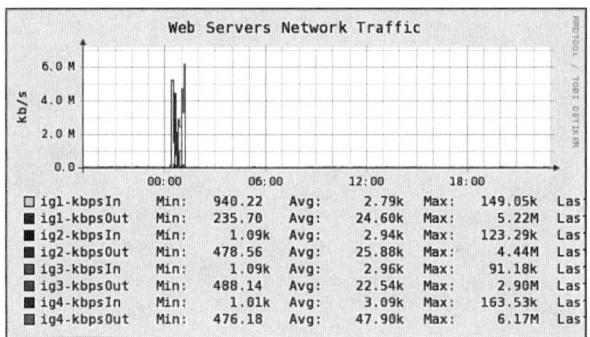

Figure 8.3   Network utilization graph, past 28 hours

Figure 8.4   Three pieces to the visualization puzzle

When packages that store data get into the business of collecting data, they sometimes make assumptions about what data you want collected and how you want it done. Likewise, when display packages get into the business of polling or storing data, they make assumptions about where the data will be and how it will be stored. This makes it difficult to bring together a patchwork of data from all kinds of devices all over the network. Packages that specialize in their singular task make fewer assumptions about what you want to do next.

There are quite a few ways to skin this cat, so I'm going to cover three visualization strategies—one each for small, medium, and large installs. But first we need to delve into the intricacies of RRDTool, because just about every time-series display solution you'll come across will be a wrapper or front-end for it. Further, RRDTool has a unique and sometimes confusing storage methodology that you must master, because the decisions you make about data storage up front impact how useful your data will be in the long run.

Enter RRDTool

RRDTool owes its lineage to the ever-popular MRTG. Written by Tobias Oetiker in 1994, MRTG has been the industry standard for graphing utilization data from network gear.

MRTG introduced the concept of the "Round Robin Database," or RRD, in the context of systems metrics collection and storage. An RRD is a circular data structure, like a ring-buffer—a database that, when full, begins overwriting itself. This structure is perfect for its intended purpose. After it is allocated, the RRD never grows in size on the disk and allows for fast data retrieval and manipulation. MRTG is of limited use for general-purpose systems monitoring because it does its own polling, but the RRD concept had great potential.

In 1999, while Tobias Oetiker was on sabbatical at the Cooperative Association for Internet Data Analysis (CAIDA), he began work on RRDTool, which is exactly what the systems-monitoring community needed for time-series data visualization. By extracting the relevant parts of MRTG into separate utilities and by extending their functionality, Tobias created a category killer for graphing time-series data.

I don't say that lightly. It's difficult to express how perfectly RRDs solve the problem of storing performance and monitoring metrics. Once stored in RRDs, the data is available via command-line tools for whatever shell processing you might want to do, but it's the fast and powerful internal processing that makes RRD a net gain.

Storing and graphing time-series data is a more difficult problem than it sounds, and RRDTool is feature rich, making it one of the most complex and difficult-to-use pieces of software that systems administrators deal with daily. It simply isn't feasible to commit RRDTool's command-line syntax to memory, so in practice, automation is the rule. Most systems administrators use scripts and glue code to hide RRDTool's complexity, and it's tricky to do this without losing flexibility. The more you know about RRDTool, the better off you are in dealing with and choosing the scripts you will use to glue RRDTool to other applications like Nagios.

RRDs are created to hold a given set of metrics. These metrics can be anything from the temperature of a room to the throughput of a router interface in bits per second. RRDTool calls these metrics "data sources" (DS). A single RRD can hold as many individual data sources as you like, but you must specify them up front when the RRD is created. Command-line tools are provided to create and periodically update the database with new values. Each data source can be stored in the database as one of several types, including GAUGE and COUNTER, which are the two types you'll most often use. To understand why different data types are needed, consider how our sample metrics, temperature, and throughput are collected.

RRD Data Types

The temperature measurement I mentioned previously will come from some sort of sensor, either inside a server with IPMI or LMSensors or from some dedicated external sensor via SNMP. When we poll the sensor, we get back a number, such as 42. This number directly corresponds to the temperature, usually in degrees Celsius. This number may increase or decrease, depending on the environment. In the physical world, a gauge with a needle pointing to a number could be used to display the temperature; hence, values of this type are normally stored as GAUGE.

Router throughput, on the other hand, might arrive via SNMP from a byte counter on the router. This number will be the total number of bytes received by the router. The counter only increments and never decreases. Depending on the size of the memory buffer used by the router manufacturer, this counter will eventually overflow, return to 0, and then continue incrementing from there. In the physical world, this counter is similar to the odometer in a car. To get an actual throughput measurement at a point in time, such as 20kb/s, we must compute the derivative for the time series. Deriving rate information from counters is such a common requirement that RRDTool will compute the derivative for you as long as you store the metric as a COUNTER or DERIVE data type.

Heartbeat and Step

In addition to the type of data source, RRDTool also needs to know how often you poll it. This number is referred to as the step. The step is specified in seconds, so in my previous example, when Nagios is polling every minute, the step would be 60. The step is not specific to the data source; it is specified once for the entire round-robin database. So the RRD expects that at least once every 60 seconds, someone will come along and use the command-line tool to give it more data. But what happens when two updates happen in a 60-second window? Or none at all? In a perfect world, our polling engine would always provide the RRD with the data it needed, exactly when it was expected, but because this is the real world, we need to account for oddities. For this reason, you must also specify a heartbeat for the data source.

The heartbeat determines what happens when there is too much or too little data. Literally, it is the maximum number of seconds that may pass before the data is considered unknown for a polling interval. The heartbeat can provide a buffer of time for late data to arrive, if set greater than the step, or it can enforce multiple data samples per polling interval, if set lower than the step. To continue our example, if the router was unreliable for some reason, or Nagios had a lot to do, we could set the heartbeat at twice the polling interval: 120. This would mean that two minutes could pass before the data for that interval was considered unknown. In practice, it's common to use a large heartbeat such as this to account for glitches in the polling engine.

Every step seconds, RRDTool takes the data it received during the step and calculates a primary data point (PDP) from the available data. If the heartbeat is larger than the step and there is at least one data sample in the `heartbeat` period, RRDTool uses that data. If there is more than one sample, RRDTool averages them. Figure 8.5 shows how RRDTool reacts to glitches when the `heartbeat` is larger than the `step`.

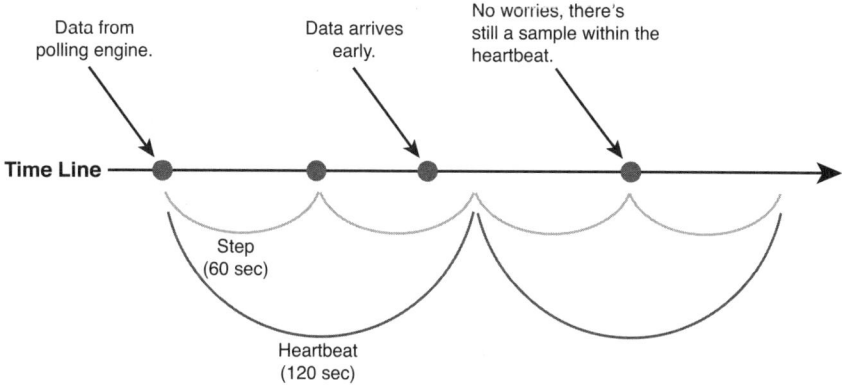

Figure 8.5 Data polling glitches with a large heartbeat

If the `heartbeat` is smaller than the `step`, multiple data samples per `step` are required to build the PDP, depending on how much smaller the `heartbeat` is. When multiple samples are required to create a PDP and enough samples exist, they are averaged into a PDP. Otherwise, the data set for that `step` is considered unknown. With a small `heartbeat`, you need more data from the polling engine than would otherwise be necessary to account for glitches. Figure 8.6 depicts a small `heartbeat` situation.

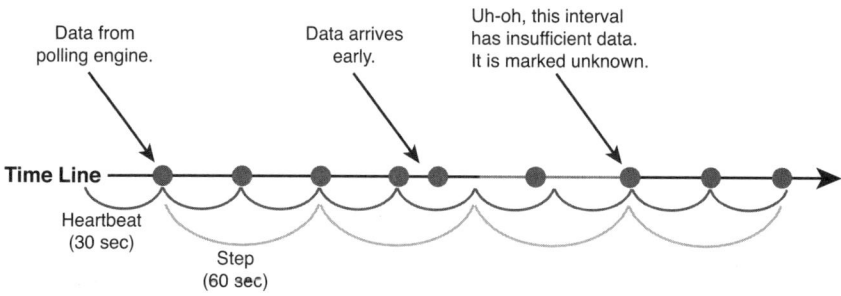

Figure 8.6 Data polling glitches with a small heartbeat

Min and Max

You may define a minimum and maximum range of acceptable values for your data source, as a sort of sanity check. If the value returned by the polling engine doesn't make sense as defined by the min and max values, the data source value for that polling interval will be considered unknown. If you do not want to specify a max or min, you may specify a U (for unknown) instead.

A single RRD can accept data from any number of data sources. Because the RRD must allocate all the space it needs up front, the data sources are defined when the RRD is created with the rrdtool create command. This means that you must decide, up front, how much data you want to keep. Because the data inside the RRD might be useful and disk space is cheap, you should keep as much of it as possible for as long as you think it might be relevant. Keeping data for more than a year is quite common in practice. Long-term data is useful for important undertakings, such as convincing management to upgrade hardware.

Round Robin Archives

If you were to poll an SNMP counter on a router every 60 seconds for a year, you would have somewhere around 525,600 primary data points (PDPs). Storing this much for every metric you monitor, on every server you monitor, quickly adds up to a lot of disk space and can slow down the creation of graphs for large data sets, so RRDTool helps you with its built-in consolidation features. You may tell RRDTool to automatically consolidate your data using a number of different built-in consolidation functions. These functions include: AVERAGE, MIN, MAX, and LAST.

For example, if we measure temperature in our cabinet and we were interested only in the maximum temperature the cabinet reaches each week, we could tell RRDTool to use the MAX consolidation function on our data every week. RRDTool then creates an archive containing just the highest weekly temperature. This archive of consolidated data is called a Round Robin Archive (RRA). You may define as many RRAs as you want, although you must specify at least one. RRAs allow us to consolidate our data in several ways in the same database. This turns out to be a great answer to the space and speed problems associated with keeping large amounts of long-term data.

For example, we might specify three different RRAs for Router7's octetsIn counter. The first might hold thirty days' worth of uncompressed data (43,200 PDPs). The second might store one year's worth of data by using the AVERAGE consolidation function to average the primary data points every ten minutes. This would keep one data point for every ten minutes (4,230 averaged PDPs). The third RRA might keep five years of data by averaging the data points every 20 minutes (10,800 averaged PDPs).

RRDTool automatically returns the highest resolution data possible. When we graph anything under one month, we get a very high-resolution graph because RRDTool returns raw data. Graphing data older than one month, but more recent than one year, yields a lower resolution graph, but more than enough data is available for us to draw the conclusions we need at that scale. Finally, low-resolution graphs are possible on five years' worth of data, while the total PDP count remains at only 58,320. That's five times as much history, at nearly an order of magnitude less storage space, than if we had kept one year's worth of raw data.

When we consolidate data in a RRA from multiple primary data points, it's possible that some of the PDPs may be unknown. So, such as with the `heartbeat`, we need to tell RRDTool how many of the PDPs can be unknown before the consolidated data point is also considered unknown. This value, called the X Files Factor,[1] is a ratio, or if you prefer, a percentage (ranging from 0 to 1) of the PDPs that can be unknown before the consolidated value is also considered unknown.

RRDTool Create Syntax

In concept, it's simple enough. DS definitions describe what and how to store, and RRAs describe how much data to keep and how often to consolidate or compress it. In practice, however, creating RRDs can be a confusing process. Listing 8.1 is the literal syntax we might use to create an RRD to hold a year's worth of data from the inOctets counter on router7.

Listing 8.1 *Creating a Single-Counter RRD*

```
rrdtool create Router7_netCounters.rrd \
--start 1157605000 --step 60 \
DS:inOctets:COUNTER:120:0:4294967296 \
RRA:AVERAGE:.5:1:43200 \
RRA:AVERAGE:.5:5:105120 \
RRA:AVERAGE:.5:10:105120
```

Yes, that's a single command, but don't be intimidated; it becomes easier the more you use it, and very few people ever bother to commit the syntax to memory. The rrdtool command works similarly to cvs. There is a single parent command (rrdtool) that operates in a number of modes (create, update, graph, and so on). So the first line puts RRDTool in create mode and tells it to name the RRD we are creating, Router7_netCounters.rrd. The second line gives RRDTool a start date and tells it to expect updates every 60 seconds. The start date is specified in UNIX's seconds-since-epoch style. You may also use an N as a shortcut for now. Specifying a start date is handy if you want to populate an RRD with old data.

In UNIX, epoch is considered to be January 1, 1970 00:00:00 UTC. Tracking seconds in this way makes working with time easier on programmers. You can use the `date` command to convert into epoch seconds, like so:

```
date +%s
```

You can convert back to a human-readable format, like so:

```
date -d "Jan 1, 1970 UTC + 1157605000 seconds"
```

Because I do a lot of this sort of thing, I find it handy to add the following line to my .bash_profile:

```
export EP='Jan 1, 1970 UTC'
```

That way, I can convert from epoch seconds back to Gregorian, like so:

```
date -d "${EP} + 1157605000 seconds"
```

Getting back to our example, line 3 defines our data source. DS definitions are colon separated. The syntax is:

```
DS:<DS NAME>:<DS TYPE>:<HEARTBEAT>:<MIN>:<MAX>
```

You may name the DS anything you want, and later you will refer to this name when you graph or export data from the RRD. The DS TYPE is either COUNTER or GAUGE, as we've already discussed. The heartbeat, minimum, and maximum values should all be self-explanatory by now. I've named our data source inOctets and specified it as a counter. The heartbeat is twice the size of the step, specified in seconds. This means that the polling engine can be late by an entire step before we consider the data for the polling interval unknown.

When you are working with counters, it's good form to set minimum and maximum values. These help RRDTool to do some internal sanity checking, thereby ensuring your hard-won data is accurately stored and depicted. My SNMP client informed me that my inOctets counter is of type 32-bit INT, so in this example, I set the minimum value to 0 and the maximum to 4,294,967,295 ($2^{32}-1$). For gauges, no math needs to be done, so if I were working with a gauge metric, I would have specified U:U.

Even if your gauge has a minimum or maximum value, I recommend that you specify U:U, so that the data is stored as it was collected. If you specify minimum/maximum for gauges, you are making a dangerous assumption about your data; namely, that it actually behaves the way you expect it should. You could lose interesting data this way. Mucking about with the data during import is bad mojo. You can perform math on the data and enforce limits later during the graphing phase.[2]

In line 4, we begin to specify RRAs. These tell RRDTool how much data we want stored and how we would like it to be consolidated. The syntax is:

```
RRA:<consolidation function>:<x-files factor>:<PDPs>:<CDP's>
```

The consolidation function will be one of MIN, MAX, LAST, or AVERAGE. The subject of consolidation functions seems to cause a lot of confusion. This is unfortunate and probably due to the fact that they are more difficult to describe than they are to understand. Simply put, the job of a consolidation function is to take a bunch of data points and to make them into a single primary data point. Each CF accomplishes this task in a slightly different but straightforward manner. Given a group of data points to consolidate, the MIN CF returns the smallest data point, the MAX CF returns the largest data point, the LAST CF returns the most recent data point, and the AVERAGE CF returns the average of all the data points.

For example, suppose I want to consolidate ten minutes of data down to a single PDP, and I am polling the data every two minutes. This means I will have five total data points, such as: 4,2,12,4,8. Given these five points, MIN would return 2, MAX would return 12, LAST would return 8, and AVERAGE would return 6.

As we discussed earlier, the X Files Factor is the number of PDPs that can be unknown before the consolidation function also returns unknown. Most people use .5 (50%) for this, but I like to go a bit higher, such as .8. At least in the context of systems monitoring, overly averaged data is better than no data at all.

In the RRDTool documentation, the next arguments—the ones I refer to as PDPs and CDPs—are called steps and rows. I'm loathe to possibly cause confusion by calling them something else, but in my experience, most people don't find these names helpful or descriptive, so forgive me for making up my own. The PDPs (or steps) argument defines how many primary data points make up a single consolidated data point. A 10 here consolidates ten PDPs into a single value, using whichever consolidation function you specify. The CDPs (or rows) argument specifies how many of these newly created consolidated data points to keep.

These two numbers, when combined with the `step`, work out to the total length of time your data will stay in the RRD before it is overwritten. For example, the `step` in our example is one minute, so a PDP value of 10 consolidates ten minutes' worth of data. Therefore, keeping five of these consolidated data points gets us 50 minutes' worth of data. A handy formula for deriving the number of days from these values is:

```
(step*PDPs*CDP's)/86400 = x days
```

That's the number of PDPs to consolidate times the number of CDPs to keep times the `step`, all over the number of seconds in a day. In practice, the polling engine usually dictates the `step` for you, and you usually know how much you want to consolidate and how many days you want to keep the data. In other words, most people need a formula for deriving the number of CDPs to keep. We can derive this from the last formula by solving for CDPs. This gets us:

```
(86400*days)/(step*PDPs)=CDPs
```

In Listing 8.1, I create three RRAs. The first effectively keeps raw data for one month. (The PDPs value of 1 says that I want to average one PDP into a CDP. The average of a single value is the value over one, so nothing is really consolidated.) I keep 43,200 of these points, so using our formula, (1*60*43200)/86400=30 days. Using the first formula, we can easily see what the other two RRAs do. The second consolidates the data into five-minute chunks and keeps one year's worth. The third consolidates even more, keeping two years of data consolidated into ten-minute chunks. For homework, create an RRA for Listing 8.1 that keeps five years' worth of one-hour chunks. (This is the type of problem you'll be solving when you work with RRDTool. The second formula makes this problem easy.)

It's possible to store as many data sources as you want in a single RRD, but new ones can't be added post-creation (easily). In Listing 8.2, I add an additional counter DS for outOctets, so we can track both the bytes coming in and going out of our router in the same RRD.

Listing 8.2 *Creating a Multicounter RRD*

```
rrdtool create Router7_netCounters.rrd \
--start 1157605000 --step 60 \
DS:inOctets:COUNTER:120:0:4294967296 \
DS:outOctets:COUNTER:180:0:4294967296 \
RRA:AVERAGE:.5:1:43200 \
RRA:AVERAGE:.5:5:105120 \
RRA:AVERAGE:.5:10:105120
```

The step in Listing 8.2 applies to both data sources, meaning that each data source you store in an RRD must have the same polling interval. However, because the `heartbeat` is specified in the data source definition, each DS may have a different `heartbeat`. In Listing 8.1, I specified 180 for outOctets's `heartbeat`. The RRAs also apply to all data sources. They will consolidate data for each of the data sources, so each data source you add to an RRD must be stored in the same manner.

After we've created an RRD, updating it is very simple. Just call RRDTool in update mode, with a filename, date stamp, and colon-separated values. RRDTool derives which values go with which data sources by matching them up in the order they were specified during creation. For example,

```
rrdtool update netCounters.rrd N:42:15842
```

updates the netCounters RRD with the current time and a value of 42 for inOctets and a value of 15,842 for outOctets.

In the years I have been working with RRDTool, I've become shy about putting a lot of data sources in a single RRD. There are two main reasons for this. The first is that I always eventually forget what data is inside an RRD. Although it's easy to derive what's stored inside an RRD by using RRDTool in fetch mode, my problem is that I forget I ever stored the data in the first place, so when I'm looking for data of type X, I'll take a cursory glance at the filenames and end up creating a new, redundant RRD with the data I want.

The second reason is that new DSs will sometimes pop up that should go in an existing RRD. For example, if you want to track the utilization of various partitions on your hard drives, you could create a single RRD called disks.rrd with a DS for each partition. Then a year later, when /var runs out of space and you add a new disk and mount it to /var, the RRD will have to be expanded to include an additional DS for the new partition. Adding DSs to an existing RRD is possible, but not fun.

My advice is to create single-DS RRDs with especially descriptive filenames. The filename of the RRD should, at a minimum, contain the name of the server it refers to. If you keep single-DS RRDs, it's also possible to put the name of the DS in the filename, which makes things especially easy if you're a blockhead like me. For example, I can look at a file called router7_inOctets.rrd and know exactly what I can get and how. The various RRDTool scripts are good at insulating you from this sort of thing, which is nice. Finding good scripts can be a problem, however, as you'll see in the next section.

RRDTool Graph Mode

Before we can get into what to look for in a graphing front-end, we need to understand a bit about RRDTools graphing mode. Like create and update, RRDTool has a special mode for drawing graphs from the data stored in an RRD. I've sometimes quipped that rrdtool graph is an existence proof of the old adage "A picture is worth a thousand words," because RRDTools graph mode is, by far, the most complex and confusing aspect of the toolset, possessing a dizzying array of options to specify everything from the height and width of a graph to its background and foreground colors.

Though graph commands are technically one-liners, it's not uncommon for them to span 15 to 20 lines. Most appear to be shell scripts. A good front-end makes the complicated syntax moot, but to get everything you can out of our FE, we need to quickly review a few important concepts, including DEFs and CDEFs, and brush you up on your reverse polish notation (I wish I were kidding).

DEFs are the core definitions in graph mode; they tell RRDTool what to graph. A DEF is made up of the filename of the RRD, a DS within that RRD, and a consolidation function for the DS. A DEF refers to one, and only one, data source. You may specify any number of DEFs in a graph command. For example:

```
DEF:foo=/usr/nagios/rrd/umbra_load:5min:AVERAGE
```

The DEF is given a name, such as a variable; in this case, foo. The foo DEF refers to the five-minute data source in the /usr/nagios/rrd/umbra_load RRD. The average keyword specifies that you want the data from RRAs that use the AVERAGE consolidation function (in case you have more than one type).

This DEF can now be graphed with a graph element definition. There are several types of graph elements; the most commonly used ones are AREA and LINE. LINE graphs look like the graphs in Figures 8.1 through 8.3. AREA graphs look like the graph in Figure 8.7.

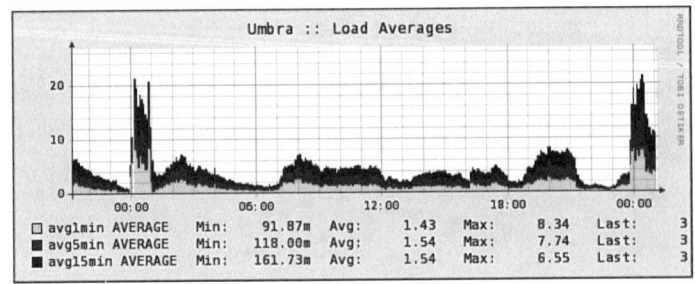

Figure 8.7   Area graph with three data sources

The only difference between a LINE and an AREA definition is the word LINE or AREA. Lines may be drawn in three thickness levels. LINE1 is the thinnest and LINE3 the thickest. The graphic element definition looks like this:

```
AREA:foo#0000FF:avg5min
```

In the preceding example, foo refers to the DEF that we specified earlier. Following the variable name foo is an RBG Hex color code and then a legend label. The combination of the DEF with the graphical element definition serves to tell RRDTool what and how to graph. Interesting things can be done, however, when CDEFs enter the picture.

A CDEF is a variable, such as a DEF, but instead of being derived directly from a DS in an existing RRD, the CDEF is derived by performing math on one or more DEFs. This real-time number crunching works transparently for the entire time series and is handy. By way of an example, check out Figure 8.8, which represents the network throughput of a web server for the past 18 hours or so.

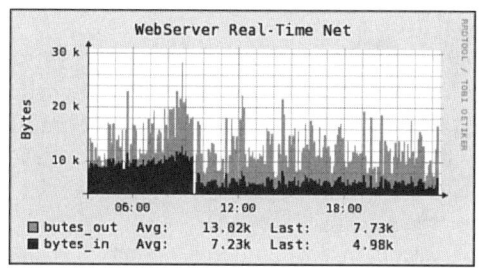

Figure 8.8   A somewhat cluttered network throughput graph

As you can see, we have two data sources: one for bytes in and one for bytes out. Although the graph is readable, it has a cluttered appearance, and it's hard to visually correlate the relationship between in and out. If we multiply the bytes_in data source by negative one, however, such as in Figure 8.9, the two counters appear on opposite sides of the Xaxis, making relationships easier to spot.

The CDEF makes this possible by creating a virtual DEF derived from operations on values from real DEFs. Listing 8.3 shows the relevant definitions in the graph mode command I used to draw the graph in Figure 8.9.

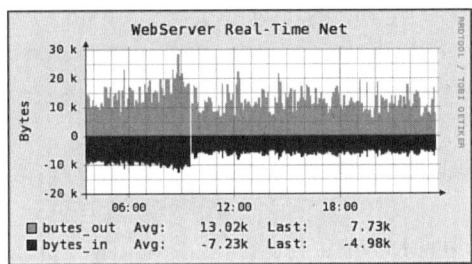

Figure 8.9   *Multiplying in_bytes by negative one*

Listing 8.3   *CDEF Syntax*

```
DEF:out=/usr/nagios/rrd/webServer_bytes_out.rrd:sum:AVERAGE \
DEF:in=/usr/nagios/rrd/webServer_bytes_in.rrd:sum:AVERAGE \
CDEF:negIn=in,-1,* \
AREA:out#FFA500:butes_out \
AREA:negIn#0000FF:bytes_in   \
```

As you can see, the negIn CDEF is a result of a math operation on the in DEF, but the expression (in,-1,*) might look odd to you, unless you've owned a fancy HP calculator or are otherwise familiar with Reverse Polish Notation. For the uninitiated, I'll provide a brief summary. Feel free to skip ahead if you are already familiar with RPN.

RPN

Traditional mathematical expressions rely on operator precedence to determine the order of operations. You may remember this from grade school as PEMDAS: Parenthesis, Exponent, Multiply, Divide, Add, and Subtract. For example:

```
4+5*2
```

The product is 14. First, 5 is multiplied by 2, and then 4 is added. This is because the operator precedence specifies that multiplication must happen before addition. If we want to multiply 2 by the sum of 4 plus 5, we must override the operator precedence with parentheses, like so:

```
(4+5)*2
```

In RPN, operator precedence is not necessary because the quantities and operators are specified in the order they are needed. For example, to specify that 5 should be multiplied by 2 in RPN, use:

```
5,2,*
```

RPN expressions are never ambiguous about the order of operations. They read from left to right, like English, and can be thought of in terms of a horizontal stack. Values are pushed onto the stack and popped off as needed. First we push a 5 onto the stack, then a 2, and then a multiplication operator. Every time RPN gets a pair of quantities onto the stack with an operator, it pops them off in order and performs the equation, saving the product back onto the top of the stack. To specify that 2 should be multiplied by the sum of 5 and 4, we could say:

```
5,4,+,2,*
```

Because RPN reads left to right, there is no chance that the multiplication could happen first. Here, 5, 4, and + are popped off the stack and evaluated. The sum of 5 and 4 is then pushed back onto the top of the stack. At this point, we can imagine the stack looking like:

```
9,2,*
```

Now, 9, 2, and * are popped off the stack and evaluated, returning 18. RPN never evaluates more than two quantities at once, so if you prefer, you may stack up all your quantities and then list out all your operators. For example, another way to write the previous equation is:

```
5,4,2,+,*
```

Here, RPN pops 5 and 4, and then, seeing that the next object in the stack is a quantity instead of an operator, skips over 2 and looks for the next available operator, which is a +. The 5 and 4 are then added and the sum is pushed back on the top of the stack. From there, execution continues in the same manner as in the last example. Some people find stacking quantities and operators such as this easier to comprehend, especially when the expressions get large.

Now that we know what RPN is all about, let's take another look at our CDEF in Listing 8.3:

```
CDEF:negIn=in,-1,*
```

Applying what we know about RPN, it's easy to see that we are creating a CDEF called negIn, which is the product of the values of in and -1. You may create a CDEF from any number of operations on any number of variables that have been defined. Let's look at a more complicated CDEF example to give you a feel for what's possible.

Listing 8.4 contains the definitions from a disk utilization graph. The RRDs track the disk metrics megabytes total and megabytes used. We are using data from four different web servers. Graphing the raw data would yield eight lines, one line for Total Megabytes and one line for Used Megabytes, for each of the four web servers.

In this example, we want to draw a graph for a presentation to management. This graph should depict the month-long history of a single, easily understandable number that quantifies the overall disk utilization on all four web servers. The best way to show a single disk metric across multiple machines is to depict the average disk utilization as a percentage. This puts all the servers on the same scale, regardless of the size of their disks. To do this, we must first convert the raw utilization metrics on each server from megabytes to percentages; then we can average the percentages into a single number. Listing 8.4 shows the RRDTool syntax to accomplish this.

Listing 8.4   *CDEFs for Data Summarization*

```
DEF:w1t=/usr/nagios/rrd/web1_disk.rrd:root_total:AVERAGE \
DEF:w1u=/usr/nagios/rrd/web1_disk.rrd:root_used:AVERAGE \
DEF:w2t=/usr/nagios/rrd/web2_disk.rrd:root_total:AVERAGE \
DEF:w2u=/usr/nagios/rrd/web2_disk.rrd:root_used:AVERAGE \
DEF:w3t=/usr/nagios/rrd/web3_disk.rrd:root_total:AVERAGE \
DEF:w3u=/usr/nagios/rrd/web3_disk.rrd:root_used:AVERAGE \
DEF:w4t=/usr/nagios/rrd/web4_disk.rrd:root_total:AVERAGE \
DEF:w4u=/usr/nagios/rrd/web4_disk.rrd:root_used:AVERAGE \
CDEF:pct1=w1u,w1t,/,100,* \
CDEF:pct2=w2u,w2t,/,100,* \
CDEF:pct3=w3u,w3t,/,100,* \
CDEF:pct4=w4u,w4t,/,100,* \
CDEF:M=pct1,pct2,+,pct3,+,pct4,+,4,/ \
```

First, DEFs are created for the requisite metrics on each server. Next, we proceed to create a utilization percentage for each server. This is done in RPN by dividing the amount of used space by the amount of total space and multiplying the result by 100, like so:

```
CDEF:pct1=w1u,w1t,/,100,*
```

After we have a utilization percentage for each server, we average the percentages into a single number by adding them up and dividing the result by four (the number of servers).

```
CDEF:M=pct1,pct2,+,pct3,+,pct4,+,4,/
```

Note that the percentage variables this CDEF operates on are CDEFs, so you can perform math on CDEF values just like you can on DEFs. If you prefer, we could use value/operator stacking to make this last CDEF a bit neater, like so:

```
CDEF:M=pct1,pct2,pct3,pct4,4,+,+,+,/
```

Data Visualization Strategies: A Tale of Three Networks

In the early paragraphs of this chapter, I said that there were three pieces to the visualization puzzle: polling engines, data storage back-ends, and graphing UIs. The way you as an administrator combine these pieces will determine the stability, scalability, and usefulness of the final solution. Myriad options are available to you, and I have no hope of covering them all. Even if I did, there would be more by the time I finished, so rather than redocumenting every last detail of the options that are currently available, I'm going to present three allegories, one each for three very different environments. In each story, the admin involved has solved his monitoring and visualization problem using Nagios along with some popular open-source data visualization tools.

Suitcorp: Nagios, NagiosGraph, and Drraw

Jason is one of three sysadmins at Suitcorp, a Fortune-500 style company with fewer than 5,000 servers. His environment is all onsite in a nine-story office building, and consists of a hodge-podge of Windows and AIX with a few Linux boxes here and there. Jason has been using Nagios to monitor system and service availability and is monitoring fewer than 20,000 services on a beefy 2-u system. The suitcorp CIO, who recently moved casual Friday to Tuesday to test whether anyone reads his email, gave Jason a healthy stipend to enhance the monitoring system to include time-series visualization—a process this month's CIO magazine said was "challenging and expensive."

Jason's monitoring box, as a result of the stipend, has plenty of cycles to spare, so Jason wants to add time-series graphs to the Nagios UI and plans to extract metrics from his existing service-check data. He's read about RRDTool and has wisely decided to use it for data-storage. All he needs is a lightweight glue-layer between Nagios and RRDTool, which will help him parse metric data from Nagios and store it in RRDs. Then he'll need an RRDTool Graphing UI to draw and present graphs for him.

This architecture, depicted in Figure 8.10, is the most common way to integrate Nagios with time-series performance visualization systems. Use Nagios as the metric poller, store the data in RRDTool, and take your pick of front-ends. The most common problem admin will run into with this architecture is choosing an inflexible glue layer between Nagios and RRDTool.

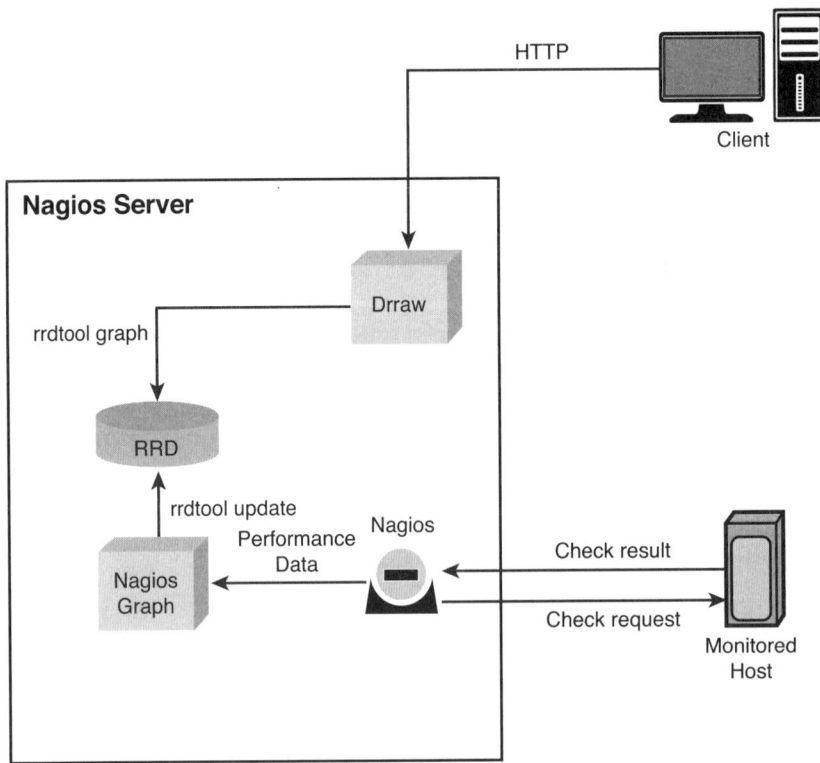

Figure 8.10   Traditional performance data graphing architecture with Nagios, NG, and Drraw

If, for example, Jason chooses a glue layer that doesn't let him define the RRAs, he won't be able to control how much data he keeps, or for how long. I make this point because very few Nagios-RRDTool glue layers allow you to specify custom RRAs in a configuration file; most of them force you to modify code. This is a big problem. Jason has a super-flexible polling engine and a perfect data storage back-end for his task, but the flexibility of both could be severely limited by the capabilities of the glue layer he puts between them.

The authors of the scripts are trying to insulate Jason from the inner workings of RRDTool, but all they've done is make his task more difficult. As a result, he must be careful in choosing a glue layer that meets his needs—one that won't become a liability later. This means picking one written in a language he understands so that he can change things like RRAs in code if he needs to, or at a minimum, making sure he can at least figure out what the RRAs are so he can plan accordingly.

Many RRDTool wrappers make the mistake of overreaching. Most of the scripts out there that collect data from Nagios and store it in RRDs also come with their own web interfaces to draw graphs from the data. In my experience, scripts that do too much tend to lock you in to doing things their way by, for example, storing the data in a manner their user interface expects. So Jason also needs to be wary of unwanted assumptions. He needs to choose a script that can be used separately from its user interface.[3]

NagiosGraph and PNP4Nagios

Through careful research, Jason has narrowed his choices down to two options: PNP4Nagios and NagiosGraph. PNP is the most popular choice as a glue layer between Nagios and RRDTool. In fact, it is the glue layer used by the commercial version of Nagios, Nagios XI. Both are lightweight, written in Perl, and easily maintained. Both have excellent integration with the Nagios UI, including, for example, icons in the Nagios status.cgi which, when moused over, pop up graphs. The first important difference between them is that PNP supports an asynchronous, bulk-update mode, whereas NG must run for every service update on every host. The second is that NG can parse metrics out of any plug-in output, whereas PNP can only read performance data output, requiring every plug-in to support performance data.

Jason isn't too worried about his monitoring system being able to handle the additional load of NG, and he doesn't relish the thought of modifying his plug-ins to support performance data, so he chooses NagiosGraph. It comes with two scripts, one for data collection and RRD storage and one for drawing HTML pages with graphs. Jason elects to use the polling/storage script and ignore the included web-UI, preferring to look for a dedicated UI instead. NG uses straight Perl regular expression syntax to define service output and autodetect new hosts. Alas, Jason cannot define RRAs in a configuration file, but grepping for "RRA" in the insert.pl program (NG's data collection script) yields the output in Listing 8.5.

Listing 8.5 *Modifying RRAs in Nagios Graph*

```
$ds  .= " RRA:AVERAGE:0.5:1:600";
$ds  .= " RRA:AVERAGE:0.5:6:700";
$ds  .= " RRA:AVERAGE:0.5:24:775";
$ds  .= " RRA:AVERAGE:0.5:288:797";
```

It's not a text configuration file, but it's close enough. Even if he wasn't familiar with Perl, Jason could modify those RRAs to get what he wanted. NagiosGraph is easy to install. There are two configuration files, one that sets up NG's defaults and one that maps the text parsed from Nagios via regular expressions into RRD variables. Simply copy the configuration files and `insert.pl` script to locations that make sense. Then edit `insert.pl` to point to the configuration files and change the configurations according to your tastes. Jason needed to change only one option: rrddir, which is the RRD storage directory. I'll come back to the map file in a moment.

Nagios graph uses the RRDTool default `step` of 300 seconds (5 minutes) and allows you to specify the `heartbeat` in its configuration file. Four RRAs are created by default, as you can see in Listing 8.5. The default RRDs are a bit conservative, in my opinion. Two days of raw data are kept, and then the average CF is applied to store 14 days of 30-minute chunks, 64 days of 2-hour chunks, and 2 years of 1-day chunks.

Nagios sends output and performance data to NG by way of the process_performance_ data directive in the nagios.cfg file. First, Jason enables performance data by setting process_ performance_data to 1 in the nagios.cfg and then adds the following line, which configures a global performance data handler:

```
service_perfdata_command=process-service-perfdata
```

Now, when Nagios gets data from a plug-in, it will execute the `process-service-perfdata` command. Listing 8.6 contains the `process-service-perfdata` command definition, which belongs in your commands.cfg or misccommands.cfg. It is this command that effectively ties Nagios to NG.

Listing 8.6 *The process-service-perfdata Command for Use with NG*

```
define command{
        command_name    process-service-perfdata
        command_line  /usr/bin/insert.pl \
  "$LASTSERVICECHECK$||$HOSTNAME$||$SERVICEDESC$||\
  $SERVICEOUTPUT$||$SERVICEPERFDATA$"
```

NG's map file contains definitions in Perl regular expression syntax. Lines passed from Nagios to NG's `insert.pl` script that match the regular expressions in the map file are parsed and made into RRDs. If no RRD for a particular service exists, NG creates it; otherwise, NG updates the existing RRD. The map file may be the thing I like best about NG. If you are adept at Perl, the flexibility inherent in this definition style is great. Listing 8.7 is a simple definition for the `check_ping` process. It matches the output of the `check_ping` plug-in and extracts the round-trip time and loss percentage, adding them both to a single RRD.

Listing 8.7 *NG's check_ping Definition*

```
#   output:PING OK - Packet loss = 0%, RTA = 0.00 ms
/output:PING.*?(\d+)%.+?([.\d]+)\sms/
and push @s, [ ping,
                [ losspct, GAUGE, $1        ],
                [ rta,     GAUGE, $2/1000 ] ];
```

This is straightforward Perl. The objective is to build a data structure referred to as an array of arrays, or AOA. The first value in the top-level array is the name of the service. This name, along with the Nagios hostname and service description, will be used in the filename of the RRD. The second and third values in the array are themselves arrays, which house the names, data types, and values of the DSs that will populate our RRD.

If you are familiar with regular expressions, you should have no problem setting up new services by copying the existing ones. Because the map file is Perl, it gives you the flexibility to accomplish some pretty neat stuff programmatically, if you have some Perl chops. For example, Jason had a problem with Disk metrics. Various boxes have different numbers of partitions, so a single static regular expression wasn't capturing all partitions on all boxes. He solved it with the definition in Listing 8.8, which dynamically detects partitions, creating a new DS for each. Pretty clever, Jason.

Listing 8.8 *A check_disk Definition for NG*

```
/perfdata:\/=\d+MB;/ and do {
  my @_perf = /(\/\w*)=(\d+)MB;(\d+);\d+;\d+;\d+/g;
  my @_s;
  my $_sref=\@_s;
  $_s[0]='disk';
  while ( my($_name,$_used,$_total) = splice @_perf,0,3 ) {
    my $_free=$_total-$_used;
    if($_name=~/^\/$/){
      $_name=~s/\//disk_root/;
    }else{
      $_name=~s/\//disk_/;
```

```
    }
    push @_s,  [ ${_name}."_total", GAUGE,  $_total*1024**2  ],
               [ ${_name}."_used",  GAUGE,  $_used*1024**2   ],
               [ ${_name}."_free",  GAUGE,  $_free*1024**2   ] ;
    }
      $s[0]=$_sref;
};
```

With the configuration files and insert.pl in place and Nagios configured accordingly, RRDs have begun appearing in Jason's rrdir. Now what to do with them?

drraw

Jason has many of the same problems choosing a UI that he did in choosing a glue layer. Most of them will not give him the fine-grained control over the DEFs and CDEFs that he wants. Eventually, he settles on drraw (web.taranis.org/drraw/) to handle data display because it is one of the very few RRDTool front-ends he can find that specializes in data display. In fact, the author clearly states that he feels it is a design flaw for a graphing engine to do things like collection and polling, and Jason couldn't agree with him more.

Drraw is composed of a configuration file and a CGI script written in Perl. It installs in a matter of minutes. Copy drraw to your CGI directory and edit it to point it at its configuration file. Again, the only thing Jason had to change in the configuration file is datadirs, which is a hash of directories where RRDs may be found. Drraw has very customizable look and feel, as well as ACL capabilities.

For Jason, drraw gets everything right. It is a lightweight wrapper around RRDTool graph, providing all the functionality in a quick, easy-to-use package. Features such as regular expression-enabled filters, templates, and graph cloning, save his wrists. Additionally, drraw does some very sysadmin-friendly things, such as storing saved configurations as text files, making it possible for Jason to create graphs and dashboards programmatically. Drraw also provides a change log so Jason can see who changed what and when.

Figure 8.11 shows drraw's home page on one of the monitoring systems at Jason's office. Drraw dashboards are collections of graphs drawn in a user-defined way. For example, you can suppress the graph legends in a dashboard or specify alternative sizes for predefined graphs.

It's not the prettiest interface out of the box, but just about everything related to aesthetics can be modified in the configuration file. This includes the icons, headers, and footers. You may even specify a custom style sheet. Figure 8.12 is a screenshot of Jason's Network Traffic dashboard.

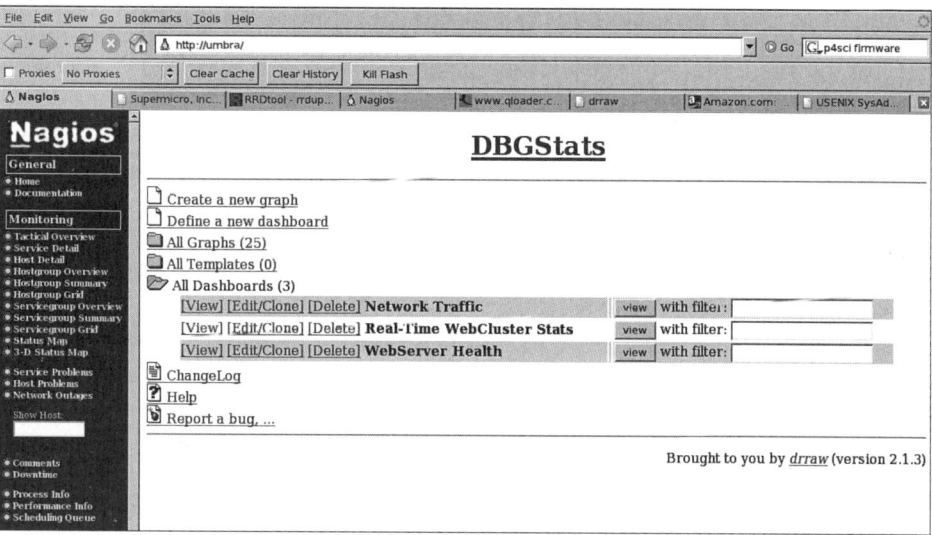

Figure 8.11 Drraw's home page

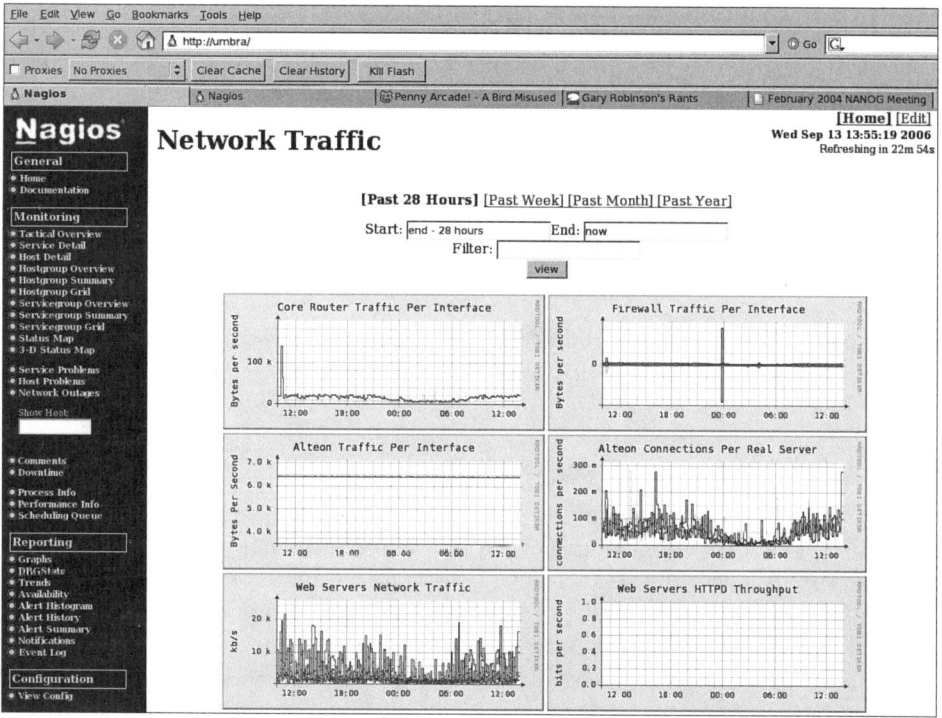

Figure 8.12 A drraw dashboard

Finally, Figure 8.13 is a screenshot of the drraw data store configuration that was used to create the consolidated disk example in Listing 8.8. You may use regular expression searches to specify RRDs and DSs for use in your graphs. Note the RRA CDEF column. This is where you can type in RPN expressions to create CDEFs.

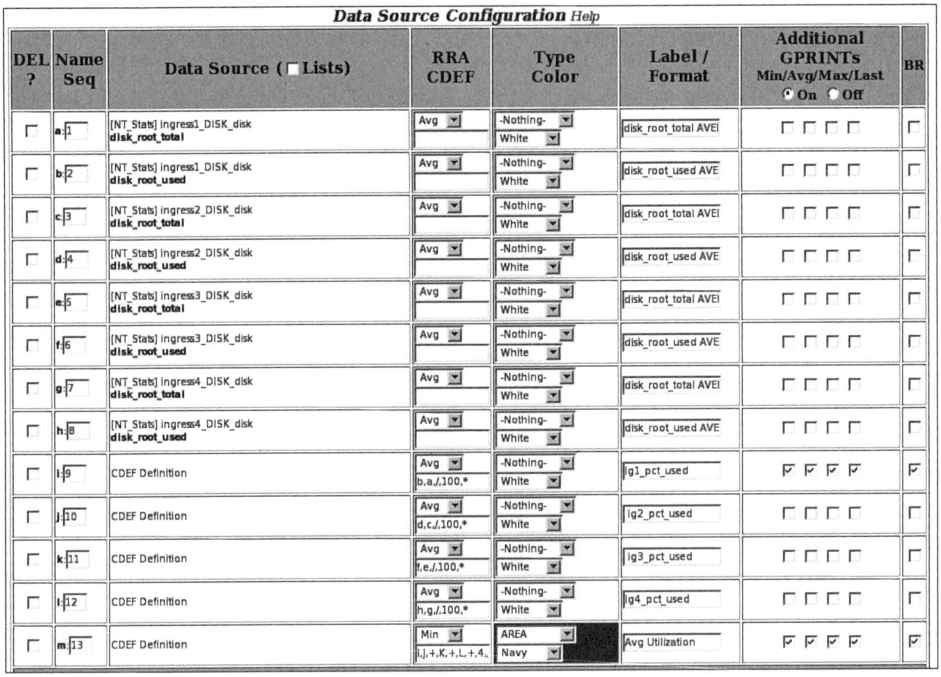

Data Source Configuration Help

DEL ?	Name Seq	Data Source (☐ Lists)	RRA CDEF	Type Color	Label / Format	Additional GPRINTs Min/Avg/Max/Last ⦿ On ○ Off	BR
☐	a:1	[NT_Stats] ingress1_DISK_disk disk_root_total	Avg ▼	-Nothing- ▼ White ▼	disk_root_total AVEi	☐ ☐ ☐ ☐	☐
☐	b:2	[NT_Stats] ingress1_DISK_disk disk_root_used	Avg ▼	-Nothing- ▼ White ▼	disk_root_used AVE	☐ ☐ ☐ ☐	☐
☐	c:3	[NT_Stats] ingress2_DISK_disk disk_root_total	Avg ▼	-Nothing- ▼ White ▼	disk_root_total AVEi	☐ ☐ ☐ ☐	☐
☐	d:4	[NT_Stats] ingress2_DISK_disk disk_root_used	Avg ▼	-Nothing- ▼ White ▼	disk_root_used AVE	☐ ☐ ☐ ☐	☐
☐	e:5	[NT_Stats] ingress3_DISK_disk disk_root_total	Avg ▼	-Nothing- ▼ White ▼	disk_root_total AVEi	☐ ☐ ☐ ☐	☐
☐	f:6	[NT_Stats] ingress3_DISK_disk disk_root_used	Avg ▼	-Nothing- ▼ White ▼	disk_root_used AVE	☐ ☐ ☐ ☐	☐
☐	g:7	[NT_Stats] ingress4_DISK_disk disk_root_total	Avg ▼	-Nothing- ▼ White ▼	disk_root_total AVEi	☐ ☐ ☐ ☐	☐
☐	h:8	[NT_Stats] ingress4_DISK_disk disk_root_used	Avg ▼	-Nothing- ▼ White ▼	disk_root_used AVE	☐ ☐ ☐ ☐	☐
☐	i:9	CDEF Definition	Avg ▼ b,a,/,100,*	-Nothing- ▼ White ▼	ig1_pct_used	☑ ☑ ☑ ☑	☑
☐	j:10	CDEF Definition	Avg ▼ d,c,/,100,*	-Nothing- ▼ White ▼	ig2_pct_used	☐ ☐ ☐ ☐	☐
☐	k:11	CDEF Definition	Avg ▼ f,e,/,100,*	-Nothing- ▼ White ▼	ig3_pct_used	☐ ☐ ☐ ☐	☐
☐	l:12	CDEF Definition	Avg ▼ h,g,/,100,*	-Nothing- ▼ White ▼	ig4_pct_used	☐ ☐ ☐ ☐	☐
☐	m:13	CDEF Definition	Min ▼ i,j,+,K,+,L,+,4,	AREA ▼ Navy ▼	Avg Utilization	☑ ☑ ☑ ☑	☑

Figure 8.13   The drraw CDEF configuration interface

With Nagios, NG, RRDTool, and drraw, Jason has all the pieces of the puzzle he needs to create great-looking time-series visualization.

singularity.gov: **Nagios and Ganglia**

Lasse is a Danish sysadmin working in Austria for singularity.gov, where an international group of super scientists compete to destroy the universe by colliding particles into each other at high speeds in the hope that they will create an antimatter explosion, or failing that, a small singularity. Lasse is the sole sysadmin for 80,000 compute nodes organized into 80 clusters of 1,000 nodes each. This is made possible by the prodigious use of automation, a configuration management system, and the fact that all the nodes are exactly the same because they work in parallel to form one big super computer created for a single, very specific purpose: particle collision analysis. All the scientists care about is timesharing pecking order and the performance of the grid as a whole.

Lasse cares how many nodes are broken at a given time, because he has to fix them. So, although he has no budget, or even any managerial reporting structure whatsoever, he managed to steal a modest box and set up a Nagios system that barely manages to perform two checks on every node every 10 minutes or so: a ping and a check_tcp on the port used by the analysis software for interprocess communication. Lasse needs to collect and graph performance statistics from every node so the scientists can see the effects of their experiments on the grid. He also needs to receive alerts when hosts run out of RAM or go above a certain CPU threshold, but he can't use Nagios as a metrics poller for several reasons:

- The Nagios Polling interval is too slow. The scientists need to take measurements every few seconds or so for the data to be useful. They want near real-time performance statistics.

- The scientists won't let him use check_ssh or NRPE because these tools noticeably impact node performance. Every CPU cycle not spent on collision analysis literally delays the destruction of the universe, and the scientists are very sensitive about that.

- Even if he could use NRPE and poll every 10 seconds, Lasse's wimpy little 1-U half-depth Nagios server wouldn't be able to handle the burden of running `rrdtool update` on the plug-in output from 160,000 service checks.

Ganglia

Ganglia is a distributed metrics collection and display system designed for high-performance parallel computing networks like the scientific clusters previously described (but that doesn't preclude its use for normal server systems). It is capable of collecting and displaying myriad performance-related metrics, including CPU, memory, and disk and network I/O in near real-time on hundreds of thousands of hosts. Ganglia accomplishes this by being a distributed application; that is, the nodes in a ganglia cluster work together to distribute the processing burden throughout the cluster. Ganglia is at least as complex a system as Nagios. In fact, entire books have been written about it, so I won't be able to do it the justice it deserves here. I can, however, give you a pretty good description and provide some integration pointers.

Ganglia consists primarily of two daemons and a web front-end written in PHP. The two daemons are called Gmond and Gmetad. Gmond may be thought of as the monitoring 'agent'. It is installed on the systems from which you want to collect statistics. Gmond is an extremely lightweight agent compared to the Nagios remote-execution programs, so Lasse's scientists will not object to its use. Each Gmond node in a ganglia cluster will multicast its metrics to the nodes in the same cluster using an efficient XDR protocol. Every node in a

Ganglia cluster knows the state of every other node in the same cluster, so the central poller needs to query only a single node from each 1000-node cluster to derive the whole cluster state.

Gmetad nodes collect statistics data from the machines running Gmond and store the data using RRDTool. Because every Gmond node knows the state of all its peers, Gmetad needs to talk only to a single node in the cluster, and it can be configured to contact back-up nodes if the node it usually talks to is down. Both daemons are written in C and make heavy use of the Apache Portable Runtime Library for portability.

Unlike Nagios, where the configuration is contained centrally, there is quite a bit of per-host configuration in Gmond's configuration file. If Lasse doesn't do too much customization, every host in the same cluster will have an identical Gmond.conf, which means he'll have 80 configuration files to maintain. This shouldn't be too much of a problem for him, because he uses a configuration management engine, but in general, this per-host configuration is about the biggest drawback to using Ganglia. In my opinion, any drawback to using Ganglia is made up for by its solid-gold web UI, Gweb, which is, bar none, the best RRDTool analysis front-end out there.

After Lasse steals another server and gets Ganglia configured, there are four ways to integrate Ganglia with Nagios. Lasse can:

- Send data from Nagios to Ganglia
- Monitor server metrics stored in Ganglia using Nagios
- Display graphs from Ganglia in the Nagios UI
- Monitor Ganglia itself using Nagios

Sending Data from Nagios to Ganglia

Gmetad generally expects to receive data from Gmond, so adding new metrics usually means writing a Gmond plug-in, but in situations like Lasse's, where he has an existing polling system (like Nagios) and just wants to share a few metrics with Gmetad, he can use the gmetric tool. Gmetric is a shell command, included with the Ganglia tarball which will inject a given metric into the data stream between Gmond and Gmetad. The most obvious way to export data from Nagios to gmetric is to use the service_perfdata_command attribute in the nagios.cfg to run a shell script that parses metrics out of the plug-in output and calls gmetric to push the metrics to Ganglia in the same way NagiosGraph and PNP4Nagios do.

The only metric Lasse wants to export from Nagios to Ganglia is the ping milliseconds from the ping plug-in output. The output from his ping plug-in looks like this:

```
PING OK - Packet loss = 0%, RTA = 0.40 ms|0;0.40
```

After enabling performance data processing, Lasse sets the following in his nagios.cfg:

```
service_perfdata_command=PushToGanglia
```

Then he defines the PushToGanglia command in his misccommands.cfg as follows:

```
define command{
command_name    pushToGanglia
command_line  /usr/local/bin/pushToGanglia.sh "$LASTSERVICECHECK$||$H
OSTNAME$||$SERVICEDESC$||$SERVICEOUTPUT$||$SERVICEPERFDATA$"
}
```

When Nagios gets the result from the check_ping service check and runs this command, replacing all the macros with real values, the input to the pushToGanglia.sh shell script will look like this:

```
1338674610||dbaHost14.foo.com||PING||PING OK - Packet loss = 0%, RTA
= 0.40 ms||0;0.40
```

I often use double-pipe characters to delimit the fields between the macros in commands like this, especially when I'm dealing with macros like SERVICEOUTPUT, because the data that is returned by service checks is itself often delimited, and there's no agreed-upon standard between plug-in writers on how the fields in a service check should be delimited. There's no telling what delimiters might already exist in the output returned by the plug-ins, so it behooves you to choose an uncommon delimiter if you're going to encapsulate data of this type. I use double pipes because two character delimiters are slightly more difficult to parse and are therefore rarely used by plug-in writers. If I absolutely must use a single character delimiter, I usually have good luck with the tilde character (~). The pushToGanglia.sh shell script looks like this:

```
while read IN
    do
    #check for output from the check_ping plug-in
    if [ "$(awk -F '[|][|]' '$3 ~ /^PING$/' <<<${IN})" ]
        then
        #this looks like check_ping output all right, parse out what we
need
        read BOX CMDNAME PERFOUT <<< $(awk -F '[|][|]' '{print $2" "$3"
"$5}'<<< ${IN})
        read PING_LOSS PING_MS <<<;$(tr ';' ' '<<<${PERFOUT})

        #Ok, we have what we need. Send it to Ganglia.
        gmetric -S ${BOX} -n ${CMDNAME} -t PING_MS -v ${PING_MS}
```

```
    gmetric -S ${BOX} -n ${CMDNAME} -t PING_LOSS -v ${PING_LOSS}

        #check for output from the check_cpu plugin
    elif [ "$(awk -F '[|][|]' '$3 ~ /^CPU$/' <<<${IN})" ]
    then
        #do the same sort of thing but with CPU data and etc…
    fi
done
```

As you can see, I'm using awk to parse the double delimiter, which is more expensive than something like cut, but, with the double-delimiter the solution is more robust.

This strategy—using Nagios performance data processing to export metrics to a script—is the "right way" to do it, but when Lasse enables these changes by HUPping his Nagios daemon, he quickly discovers that his little system can't keep up. The problem is that performance data processing is a global setting in Nagios. Every service check calls his shell script, including the ones he doesn't care about, like the check_tcp plug-in. As a result, the pushToGanglia.sh script is running twice as often as Lasse really wants it to. He needs a way to selectively enable performance data processing for just the plug-ins he's interested in.

Unfortunately, Nagios doesn't have an option to do that, but Lasse can achieve the same effect by writing a wrapper for his check_ping plug-in. I described plug-in wrappers in Chapter 2, but the general idea is to write a shell script that calls the real plug-in, captures its output, and sends the metric data to Ganglia with gmetric, before exiting in the same way the real plug-in would. Here's a wrapper for check_ping that does what Lasse needs:

```
#!/bin/sh

ORIG_PLUGIN='/usr/libexec/check_ping_orig'

#get the target host from the H option
while getopts "H:" opt
do
        if [ "${opt}" == 'H' ]
        then
                BOX=${OPTARG}
        fi
done

#run the original plug-in with the given options, 0and capture its
output
OOUT=$(${ORIG_PLUGIN} $@)
OEXIT=$?

#parse out the perfdata we need
read PING_LOSS PING_MS <<< $(echo ${OOUT} | cut -d\| -f2 | tr ";" "
")
```

```
#send the metrics to Ganglia
gmetric -S ${BOX} -n ${CMDNAME} -t PING_MS -v ${PING_MS}
gmetric -S ${BOX} -n ${CMDNAME} -t PING_LOSS -v ${PING_LOSS}

#mimic the original plug-in's output back to Nagios
echo "${OOUT}"
exit ${OEXIT}
```

The wrapper approach takes a huge burden off of Lasse's Nagios server. With wrappers, he can choose to process performance data on a service-by-service basis, or he can even selectively export ping data from some hosts and not others by defining something like check_ ping_orig and directing some hosts to use it instead of check_ping in the service definition. The biggest problem with the wrapper technique is that it's easy to forget it's there. Any upgrade or reinstall of the nagios-plugins tarball will break the data sharing arrangement. If you work in a shop with other administrators, take care to ensure that everyone is aware that plug-in wrappers are in use, so no one accidently overwrites them.

Monitor Ganglia Metrics Using Nagios

Lasse's requirement to send alerts based on CPU-thresholds can be satisfied by one of the five Nagios plug-ins that are included in Ganglia's Gweb tarball. These plug-ins may be copied into the plug-ins directory of your Nagios servers and can be run locally. Each plug-in interacts via HTTP with a series of Gweb PHP scripts that were created expressly for the purpose. The check_host_regex.sh plug-in, for example, interacts with the PHP script:

```
http://your.gweb.box/nagios/check_host_regex.php
```

Each PHP script takes the arguments passed to the plug-in and parses a cached copy of an XML dump of the grid state obtained from Gmetad. With the plug-ins provided by the Ganglia project, Lasse can:

- Check the heartbeat of a given host (analogous to a ping)
- Check a single metric on a single host (like CPU_LOAD)
- Check multiple metrics on a single host
- Check one or more metrics on various hosts (defined by a regex)
- Verify that a metric value is the same on various hosts (defined by a regex)

The check_ganglia_metric plug-in, which compares a single metric on a given host against a Nagios threshold, meets Lasse's requirement exactly. To use it, he first copies it from the Nagios subdirectory in the Gweb tarball to his Nagios plug-ins directory. He then edits it to change the GANGLIA_URL variable in the script to match the URL of his Gweb server. By default, the URL variable is set to:

```
GANGLIA_URL="http://localhost/ganglia2/nagios/check_metric.php"
```

Next, Lasse defines the check command in his Nagios commands.cfg, like so:

```
define command {
  command_name  check_ganglia_metric
  command_line  $USER1$/check_ganglia_metric.sh host=$HOSTADDRESS$
metric_name=$ARG1$ operator=$ARG2$ critical_value=$ARG3$
}
```

Lasse wants to be alerted when the 5-minute load average of any host goes above 5, so he defines a CPU service for all his hosts with the following check_command directive:

```
check_command                        check_ganglia_metric!load_five!more!5
```

Displaying Graphs from Ganglia in the Nagios UI

Ganglia's web front-end, Gweb, comes with a magical PHP script called `graph.php`. The `graph.php` script can graph any combination of hosts and attributes stored by Gmetad, given a few simple attributes in the URL. Lasse can combine `graph.php` and the `action_url` attribute in his Nagios service definitions when he wants to display graphs from Ganglia in the Nagios UI. When specified, the `action_url` attribute creates a small icon in the Nagios UI next to the host or service name to which it corresponds. If a user clicks this icon, the UI will direct the user to the URL specified by the `action_url` attribute for that particular object.

Given the uniformity of Lasse's environment, it turns out that his host and service names happen to be consistent in both Nagios and Ganglia, so it's pretty simple for him to point any service's `action_url` back to Ganglia's `graph.php` using built-in Nagios macros, so that when a scientist clicks the `action_url` icon for that service in the Nagios UI, he or she is presented with a graph of that service's metric data. For example, this is a Gweb URL that will display the graph of the ping_ms on host h1:

```
http://lasses.ganglia.box/graph.php?c=cluster1&h=h1&m=ping_
➥ms&r=hour&z=large
```

You may have noticed that Ganglia's `graph.php` requires a `c=` attribute. This must be set to the name of the cluster to which the given host belongs. The problem is that Nagios has no concept of Ganglia clusters, but Lasse can work around this using Nagios custom variables. Custom variables must begin with an underscore and are available as macros in any context a built-in macro is.

Here's an example of a custom variable in a host object definition defining the Ganglia cluster name to which the host belongs:

```
define host{
host_name       h1
address         192.168.1.1
_ganglia_cluster  cluster1
...
}
```

If Lasse had a more complicated environment, he might need to use custom variables to correct differences between the Nagios and Ganglia namespaces, creating, for example, a `ganglia_service_name` macro in the service definition to map a service called CPU in Nagios, to a metric called `load_one` in Ganglia. To enable the `action_url` attribute, Lasse creates a template for the Ganglia `action_url` like so:

```
define service {
    name       ganglia-service-graph
    action_url http://lasses.ganglia.host/ganglia/graph.php?c=$_
GANGLIA_CLUSTER$&h=$HOSTNAME$&m=$SERVICEDESC$&r=hour&z=large
    register   0
}
```

This makes it easy for him to toggle the `action_url` graph for some services but not others by simply including `use ganglia-service-graph` in the definition of the services that he wants to graph. As you can see, Lasse's `action_url` combines the custom-made `_ganglia_cluster` macro he defined in the host object with Nagios's `hostname` and `servicedesc` built-in macros. If the Nagios service name was not the same as the Ganglia metric name, Lasse would have defined his own `_ganglia_service_name` variable in the service definition and referred to that macro in the `action_url` instead of the `servicedesc` built-in.

The Nagios UI also supports custom CGI headers and footers, which make it possible to accomplish rollover pop-ups of the `action_url` icon containing graphs from the Ganglia `graph.php`. This requires some custom development, and is outside the scope of our little example, but I wanted you to know it's there. You can read more about it here:

`http://nagios.sourceforge.net/docs/3_0/cgiincludes.html`

With Ganglia, Lasse was able to completely offload the burden of performance metrics collection, storage, and display, and he even picked up a few lightweight service checks in the bargain, all while maintaining the Nagios UI as the primary interface for network and system state information.

Massive Ginormic: Nagios, Logsurfer, Graphite, and Life After RRDTool

Ted is one of umpteen-thousand sysadmins at Massive Ginormic, a huge Internet company whose billion-dollar revenue stream is a mystery to everyone, including their shareholders. Massive Ginormic employs an army of sysadmins, software developers, and recruiters, whose salaries are computed as a function of the number of Ph.D.s they hire per week. At Massive Ginormic, there are so many servers that the sysadmins quantify them as a multiple of Avogadro's number, and the network is so distributed that entire data centers have literally burned to the ground without anyone noticing for weeks.

Traditionally, internal groups at Massive Ginormic have formed ad hoc techno-fiefdoms led by individual Ph.D. warlords, or small groups of Ph.D. oligarchs. Each fiefdom was roughly self-contained, claiming some territory in the form of production servers and running as much of its own infrastructure (including monitoring systems) as possible. Recently, however, the developers, under the banner of "DevOps" and "Continuous Integration" have staged a companywide coup, and, laying siege to most of the feudal territory, have largely torn down the walls of access-control between production servers, enabling any developer to change any application, anytime, anywhere.

A renaissance of application development productivity has blossomed as a result of the coup, enabling Massive Ginormic to grow in surprising and innovative directions at a breakneck pace. This has come at the expense of hundreds of little integration bugs and race conditions in the application code, many of which go unnoticed for weeks before the customers begin to complain. If they want to continue to operate with a minimum of production control, the developers realize that they're going to need a way to visualize the effect their code changes are having on the production systems.

To this end, they have approached Ted in the hope that he can come up with a way to pull together data from the hundreds of scattered Nagios systems, as well as other kinds of metrics, including those from the code-deployment servers and instrumentation the developers have embedded in the applications themselves. The goal is to be able to draw graphs that can correlate things like the CPU utilization from a subset of servers, along with the day/time of a code deployment, and perhaps even the number of times a certain subroutine gets called within the application itself.

It doesn't take Ted long to realize that the developers have presented a data-storage and visualization dilemma that RRDTool cannot solve. The first problem is that the developers want to visualize metrics like code promotion occurrences, which do not occur on a regular interval. There is no way to set the `heartbeat` and `step` for a metric like this, and, as a result, the preponderant quantity of data points will be stored by RRDTool as "NAN" and the RRAs will eventually combine the NANs in such a way as to delete the actual data. Ted needs a data storage layer that doesn't care when data arrives, whether it arrives in a temporal order, or that it arrives consistently.

The second problem is that if the developers want to be able to instrument their software in an ad hoc way to send metrics to the visualization system, they'll be adding new types of metrics constantly, possibly on the order of thousands per day. RRDTool is very rigid about the creation of the RRDs themselves. As you've read, database creation is complex and requires an intimate knowledge of things such as `heartbeat`, `step`, and data type; most sysadmins rely on scripts to do it for them, and those scripts rely on very specific regular expressions, provided by the admin to choose the correct data type (`gauge` or `counter`, for example). Ted needs a data storage layer that is agnostic to data type; he needs storage to be allocated on-the-fly without human intervention of any kind.

The third problem is the scale. Given a central metric-collection and visualization system, Ted can easily imagine millions of metrics per minute being thrown at it. `rrdtool update` usually writes every metric to disk when it's called, unless `rrdcached` is employed, but even if Ted used `rrdcached`, the disk I/O load will quickly become too much for a single system to handle. Ted needs a distributed solution—something that he can scale.

Luckily for Ted, some systems engineers at Orbitz.com had a very similar problem and created a program called Graphite to solve it. The name refers to a suite of three discrete but complementary Python programs, one of which is itself called Graphite. (I assume they did this to make it more difficult to write books about.)

The first of these is Whisper, a reimplementation of the RRD format that makes the modifications to the data layer I mentioned previously. Whisper does not particularly care how far apart Ted's data points are spaced, or indeed if they arrive in sequential order. It also does not care what kind of data it is. Internally, Whisper stores all values in the same way RRDTool would store a `GAUGE` data point. "What about counters?" you ask. Data interpretation is handled by the front-end using various built-in functions that modify the data when it's displayed. For example, at display time, the user applies the `derive` function to obtain a bytes-per-second graph from byte counter data stored by Whisper in its raw format.

The critically important upshot is that by making the storage layer agnostic to data type and frequency, new Whisper databases can be created on-the-fly with very little preconfiguration. Ted needs only to specify a default storage configuration (and, optionally, more specific configurations for metrics matching more specific patterns), and after that, all Whisper needs to record a data point is a name, a value, and a date stamp.

Carbon, the second Python program, listens to the network for name/value/date-stamp tuples and records them to Whisper RRDs. Carbon can create Whisper DBs for named metrics that it has never heard of, and begin storing those metrics immediately. Metric names are hierarchal from left to right and use dots as field separators. For example, given the name "appliances.breakroom.coffee.pot1.temp," Carbon will create a Whisper DB called 'temp' in the $WHISPER_STORAGE/appliances/breakroom/coffee/pot1 directory on the Graphite server. Carbon listens on TCP port 2003 for a string of the form "name value date." Dates are in EPOCH seconds. Continuing the coffee pot example, Ted could update that metric with the value '105' with the following command line:

```
echo "appliances.breakroom.coffee.pot1.temp 105 1316996698" | nc -c
<IP> 2003
```

I passed "-c" to netcat so that it wouldn't block waiting for a TCP reply from Carbon, which is a "fire and forget" style daemon. Obviously you need the netcat with -c support to do that (http://netcat.sourceforge.net). Carbon gives Ted's developers exactly what they need: a socket on the network to which anyone can send data and have it stored and ready for graphing immediately. Further, Carbon may be configured to send data to multiple back-end Whisper servers, so Ted can scale the solution as large as he needs.

Whisper's data agnosticism and Carbon's network presence combine in such a way that data collection and presentation is no longer an OPS-specific endeavor. Carbon clients have been written for about every popular programming language out there, making it trivial for the developers instrument their applications to send interesting metrics to the Graphite server. For that matter, there's no reason why the security guys couldn't blast over their IDS metrics, or the recruiters their Ph.D. recruitment stats.

Finally, Graphite is the web front-end to... well, Graphite. Graphite runs on the Apache web server with mod_python and includes a novel web-based command-line interface with tab completion and wildcard support, which makes it easy to create on-the-fly graphs from any combination of stored metrics. It also has a tree view reminiscent of Cacti, and a user-configurable dashboard view. The front-end also features the URL interface, which allows the creation of graphs by specifying URLs and enables integration with Nagios via the "action_url" attribute as described previously.

Because Ted has a lot of Nagios servers spread throughout the network, and none of them are working very hard, he decides to enable performance data processing on all of them (as described earlier) to export the system metrics to Graphite. Because there is no off-the-shelf glue layer like NagiosGraph to tie Nagios to Graphite, Ted decides to write the performance data to a log and then use Logsurfer to parse the log file. Logsurfer is an open-source, real-time log parser. It's written in C and can be configured to take action on patterns in log files that match regex's that he provides. His first step is to configure Nagios to write performance data to a file, and to that end, he defines the following "process_performance_data" command:

```
command_line    /usr/bin/printf "%b\n" \ "$LASTSERVICECHECK$||$HOSTNA
ME$||$HOSTNOTES$||$SERVICEDESC$||\
$SERVICEOUTPUT$||$SERVICEPERFDATA$" >> /var/log/nagios/service-
perfdata.out
```

Then he writes a simple little shell script that uses netcat to push metrics to Carbon:

```
#!/bin/sh
SERVER=$4
PORT=$5
[ "${SERVER}" ] || SERVER=graphitebox
[ "${PORT}" ] || PORT=2003
#send data to carbon
echo "$1 $2 $3" | /usr/bin/nc -c $SERVER $PORT
```

Finally, he installs logsurfer and creates a configuration file that parses metrics from the logfile and calls his shell script to push the data to Carbon. Here, for example, is his logsurfer config for the Nagios check_ping plug-in output:

```
'^([^:]+)::([^:]+)::([^:]+)::([^:]+)::PING [A-Z]+ - Packet loss =
([0-9]+)%, RTA - ([^ ]+) ms::.*$' - - - 0 continue
        exec "/usr/local/bin/gclient.sh $4.$3.$5-loss $6 $2
graphitebox 2003"
```

Because of the architectural changes Graphite makes in the data storage layer, it is quite a departure from the tools I've covered so far. The first big difference is that, because of the imposition of Carbon with its data-caching layer between Whisper and the UI, there isn't a command-line tool for creating graphs directly from Whisper databases. You must use the web front-end. In practice, this is not really a problem. It's easy to script the creation of a graph in shell, with curl or wget running against the graphite web-UI. Given the myriad ways to export Graphite data into other systems, in general the data I place in Graphite always "feels" more accessible.

The next big difference is rooted in the fact that Graphite stores the data you give it in its raw format. The most visible example of this difference is in the computation of rate information from counter metrics. While RRDTool does this for you automatically, in Graphite, you need to apply the 'derive()' function to the raw counter data when you graph it. This may seem counter-intuitive until you become familiar with the rich selection of functions that are available.

Graphite functions may be applied in the Graphite web-UI via the Graph Data button in the Graph composer, but after you become familiar with Graphite, I suspect you'll find it more expedient to work with the URLs directly. This is both because it's faster to type function names than it is to hunt for them in menus, and also because the function names in the documentation don't exactly match those in the GUI; we avoid the need to hunt around in menus by simply typing them into the URLs. To get started, populate some data into a graph using the graph composer, then right-click the graph itself and select Copy Link Location. Finally, paste the URL into a new browser tab.

Functions apply in a C-like manner, and most of them accept multiple metrics and even wildcards in lieu of lists. Taking our earlier derive example, we can apply the function to 'some.counter.data' like so:

```
&target=derive(some.counter.data)
```

This function yields rate data as depicted in Figure 8.14, which brings up an interesting point—namely, that looking at raw counter data is sometimes interesting. This probably wouldn't have occurred to us using RRDTool, but what if we compare the raw byte-counters of two different routers, as shown in Figure 8.15?

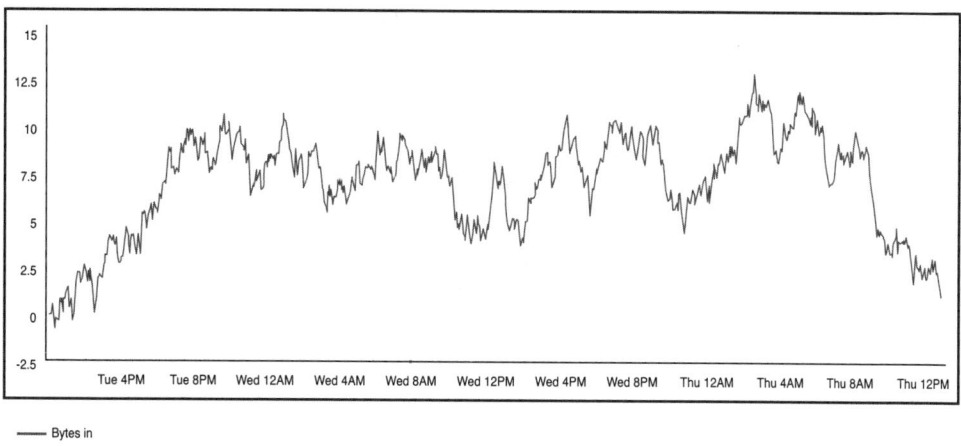

Figure 8.14   Bytes in Graph, as drawn by Graphite using the derive() function

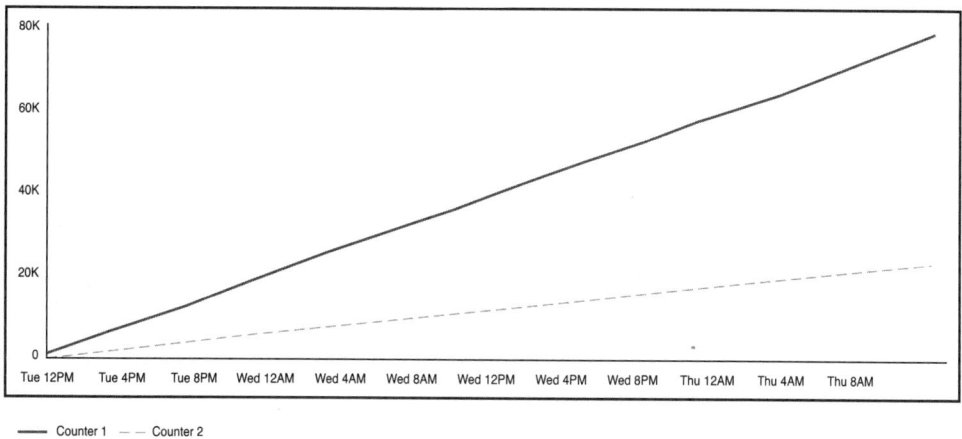

Figure 8.15   Raw counter data from two routers

This could be useful capacity planning info, but it's not a fair comparison because the routers have different total values, so one router will always appear to be growing at a smaller rate than the other. That's okay; Graphite provides us a 'secondYAxis()' function, which easily allows us to draw one of these two data sets on its own Y axis. So by graphing

```
&target=router1.bytes&target=secondYAxis(router2.bytes)
```

we can get a clearer picture of comparative rate of growth of the byte counters for these two routers, as seen in Figure 8.16. There's also an 'integral()' function, which allows you to take GAUGE-based data sources and get counter-style data. If, for example, you had a graph of widget sales per minute, you could apply the integral function to graph total sales for a given time interval.

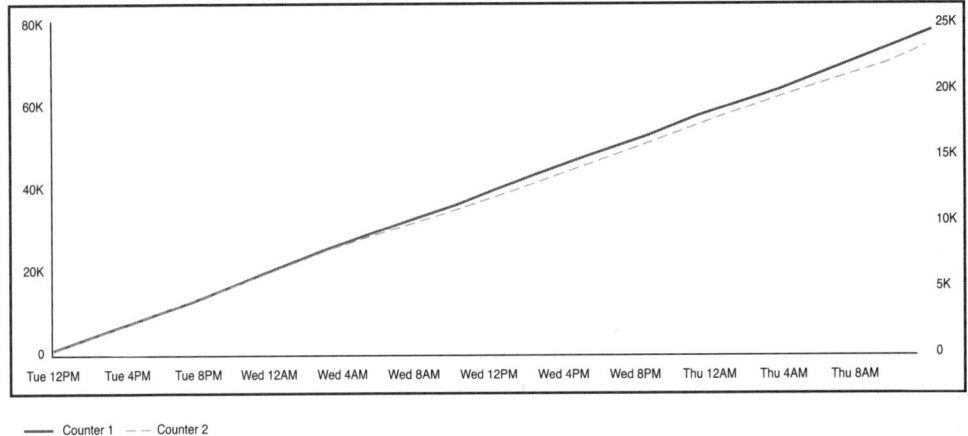

Figure 8.16   Raw data redrawn with the secondYAxis() function

If you read the previous section on RRDTool, take a moment and think about what it would have taken to draw Figure 8.16, using `rrdtool graph` especially if you had been storing your counter data as type COUNTER like you should. It's probably possible, but I admit I don't know how to do it, and the prospect of puzzling it out in reverse polish notation somehow stops short of sounding appealing. Even if I did tease it out, I wouldn't be likely to apply the technique to other data sets for the benefit of my own curiosity, and the various RRDTool-based front-ends out there wouldn't be much help to me in that endeavor. Graphite's functions do an excellent job of inviting you to visualize your data in new ways. After you get used to them, I think you'll find they are easily accessible and fun.

The functions themselves are fully documented online, and I can't cover all of them here, but let's take a look at some of my favorites, starting with `summarize`. This function allows you to recompute the interval for a given set of time-series data. So given a metric like the number of users registering for an online service, as depicted in Figure 8.17, we can imagine that the marketing team has a goal to maintain X registrations per hour and would like to display this data on a kiosk in the hallway, but they'll want it graphed as "registrations per hour" to reflect their goal. We can compute this graph, depicted in Figure 8.18, for them with the following:

```
&target=summarize(user.registrations,"1h")
```

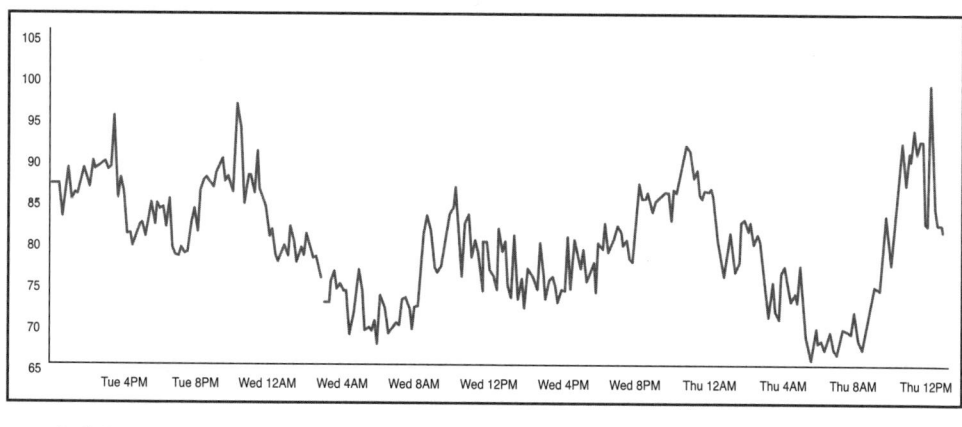

Figure 8.17 User registrations per second

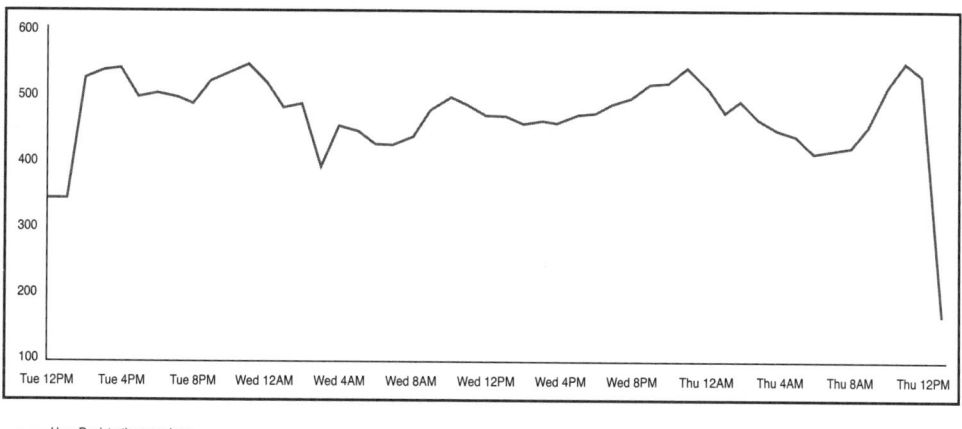

Figure 8.18 User registrations per hour, computed using the summarize() function

To make their progress more obvious, we could add a horizontal line (constant) equal to their goal (see Figure 8.19) with the threshold function, like so:

```
&target=summarize(user.registrations,"1h")&target=threshold(400,
➥"Goal")
```

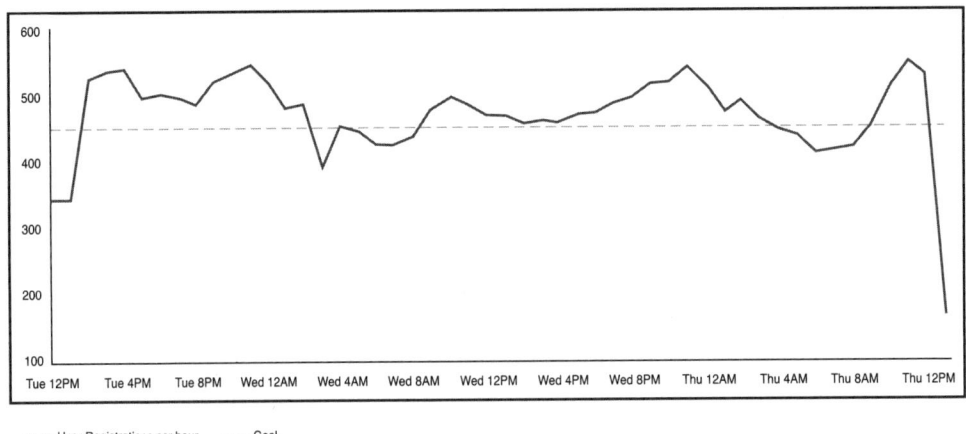

Figure 8.19   User registrations per hour with a goal constant added using the constant() function

Functions are nestable, just like in C, so we could add the data from last week to the graph by nesting the summarized target inside a `timeShift` function. This would give the marketeers some historical registration data from last week for context while still maintaining a normal period on the X-axis. This graph, drawn with the targets listed next, is depicted in Figure 8.20.

```
&target=summarize(user.registrations,"1h")&target=timeShift(summarize
(user.registrations,"1h"),"7d")&target=threshold(400,"Goal")
```

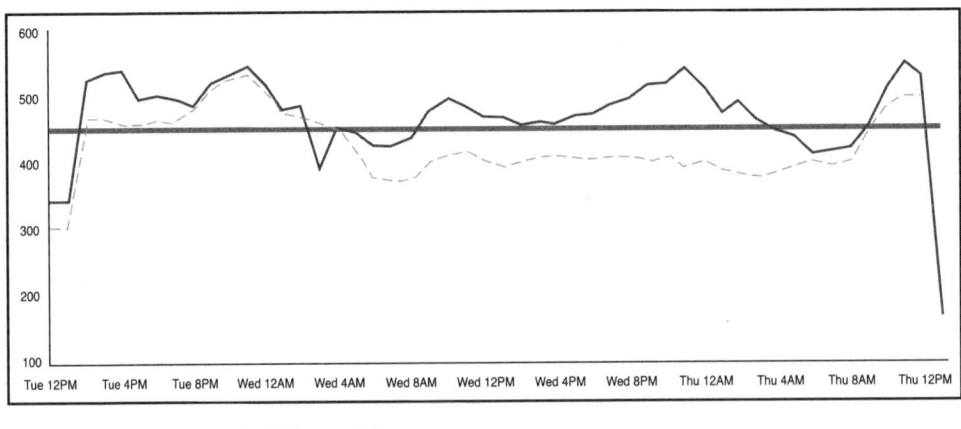

Figure 8.20   User registrations with time-shifted data for context using the timeShift() function

Various functions exist for combining multiple metrics into a single line. These are `SumSeries`, which creates a single line from multiple metrics by adding them together; `AverageSeries`, which averages multiple metrics into a single metric; `MinSeries`, and maxSeries, which plot only the minimum or maximum value data points in the series. All these functions support wildcards in the data-source field. For example:

```
&target=averageSeries(dc4.web.*.cpu)
```

plots a single line with the average CPU utilization of every web server in dc4. Combinatorial functions are great for summarizing clusters, or even data centers. I find myself combining multiple `averageSeries` of different metric types (CPU and disk for example) using `secondAxis`. In this way, I can get multiple metrics across entire data centers on the same graph in a really usable way. Other functions exist for filtering individual metrics out of large lists. For example:

```
&target=highestCurrent(dc4.web.*.cpu,5)
```

plots the CPU utilization of only the five currently most utilized web servers in DC4. Combining these:

```
&target=averageSeries(highestCurrent(dc4.web.*.cpu,5))
```

plots the average CPU utilization of the five currently most utilized web servers in DC4. These are awesome for dashboards where you want to show things that are misbehaving, or aberrant behavior in general. I'm sure you get the idea by now. The filters are too numerous to list here, but included are filters for highest and lowest max, average and current filters that plot metrics that fall above or below static thresholds as measured by max min and average, and filters to plot metrics that are most deviant from the rest of the series.

Armed with functions like these, Ted's developers can be very introspective about the changes they are making to the environment. Quite simply, the more data they throw into Graphite, the more questions they can ask.

DIY Dashboards

In the preceding sections, we've seen Nagios in a few different roles, first as an all-encompassing data collection and visualization system, then in a peer relationship with a special-purpose performance analysis tool, and finally, as one of many sensors in a large data collection and analysis architecture. In every example, however, I've been talking about time-series data. This section is here to provide you some options for data visualization beyond time-series graphs, and also to provide some advice in the context of building dashboards to suit your own needs, using some interesting tools.

Know What You're Doing

Before you start designing custom interfaces and dashboards, I highly recommend that you read up on what good data visualization is and especially what it isn't. There are many ways to get it wrong and only a few to get it right. An excellent book is Edward Tufte's *The Visual Display of Quantitative Information*. I'll give you a few tips to get you started, but I'm certainly no expert, so make sure to do your homework before you get started.

First, avoid pie charts. Although adored by marketing folks, pie charts are awful for data visualization for several reasons. Primary among these is that humans are notoriously bad at interpreting angles (3D effects on pie charts only make things worse). Given two similar values in a pie chart, most people won't register the difference unless you point it out to them or publish values in the legend labels. Second, pie charts have no frame of reference, which means that people can glean the actual value each slice represents only by reading it in a label. Finally, pie charts have no obvious beginning or end, which makes it difficult for people to know where to focus their attention.

Anything you can visualize with a pie chart is better visualized with a bar or point chart. These provide a frame of reference on two axes and communicate data more effectively to humans in general. If you absolutely have to use a pie chart, do not compare more than a few values, and make sure to colorize in such a way that the slices have a high contrast.

Bar charts are especially good at depicting categorical information, so if you have information that fits nicely into categories, use a bar chart and arrange the data so that the largest values are furthest to the left. Point charts or line charts are better for depicting time-series data. The time-series data graphed in Figures 8.21 and 8.22 give a good example of how much clearer point graphs are when information is depicted as a function of time.

Try to think about your graphics in terms of maximizing for information per pixel. You know you are doing things correctly if you are packing large amounts of data into small areas. Fancy effects, such as 3D, should be avoided for this reason. 3D effects usually take up more space and rarely add useful information. They also have a habit of obfuscating the useful information that remains. Silly widgets, such as thermometers, speedometers, or any other "ometers," should be avoided for the same reason. These things take up a lot of space to communicate a single value.

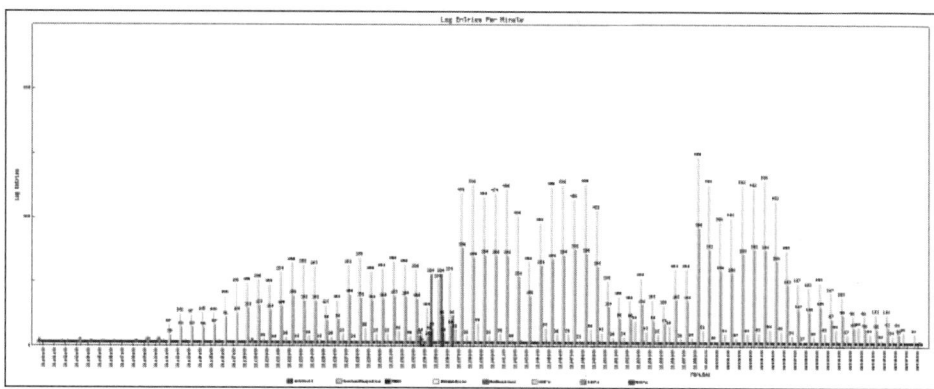

Figure 8.21   Bar charts don't depict time-series data well.

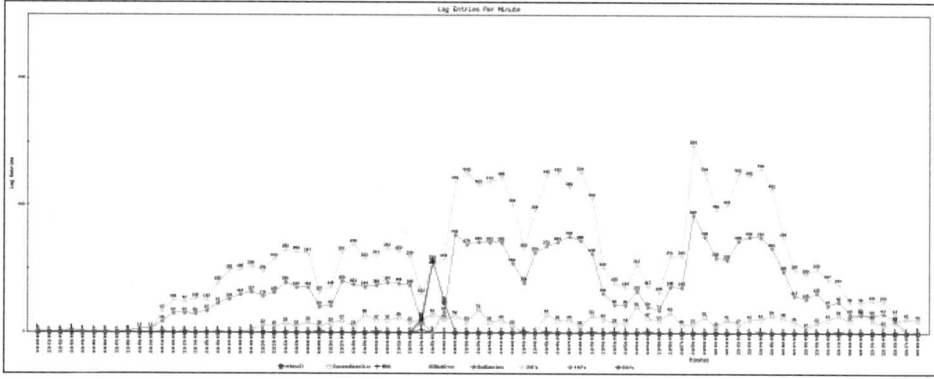

Figure 8.22   Point charts are much cleaner when time is involved.

You should attempt to keep your graphs as uncluttered as possible so that they stand on their own without requiring extra explanation. If the graphs are large and there is space, label the individual points for clarity. Axis labels should convey the units they express, and titles should be carefully thought out, short, and descriptive. Speaking of Tufte, a few graphing libraries make explicit use of his guidelines; the TufteGraph JavaScript library is one notable example.

All this highbrow advice goes out the window when your marching orders are to provide whiz-bang displays with little or no substance. Many a well-built monitoring system has been scrapped in favor of a kludgy black box with neat-o graphics that impressed the execs.

Because the well being of your hard-won monitoring solution may one day rest on your ability to come up with something other than line graphs, my last piece of advice to you is not to be a visualization elitist. Don't be afraid to throw a speedometer or two around, if for no other reason than to prove you can. By all means, guide and educate the people who will use the dashboard, but always remember that the data consumer defines the management interface.

Some of the tools that follow can provide the eye candy so coveted by the upper echelons, but all of them[4] can also be used to create interesting visual displays that are useful in a pragmatic sense. The art, or science, of data visualization is growing very quickly, especially in the area of IT security, where visualization may be the only hope for good real-time, enterprisewide event correlation in the absence of true AI. The tools referenced herein should not be considered a comprehensive list. They represent only the tip of the iceberg; by the time you read this, chances are the iceberg will have grown considerably.

RRDTool Fetch Mode

Just when you thought you were done with RRDTool, we must make one last foray. Because much of the historical data we need to visualize resides in our RRDs, we need a way to query and extract it. RRDTool, in fetch mode, can give you raw data dumps for a given period of time. To use it, provide RRDTool the name of the RRD, the CF you are interested in, and the time range in epoch seconds. For example, to get the last ten minutes of data from the server1_load.rrd, we could type:

```
rrdtool fetch server1_load.rrd AVERAGE -s 'date -d "10 minutes ago"
+%s' -e N
```

Listing 8.9 contains the output from this fetch command. As you can see, the data from fetch mode will require some processing.

Listing 8.9   *Output from rrdtool Fetch Command*

```
          avg1min              avg5min              avg15min
1158207000: 1.9546666667e-01 1.1120000000e-01 3.4666666667e-02
1158207300: 7.8000000000e-02 1.0040000000e-01 4.5066666667e-02
1158207600: nan nan nan
```

Two polling intervals have happened in the last ten minutes. Depending on when you launch the command, the last polling interval (N) may not have happened yet, so these values are nan'd (Not A Number). The values are in scientific notation. We can use the bc math language in shell to convert these values to numbers that we can use with visualization tools. Listing 8.10 is the shell script I use to process and extract the data from a fetch command.

Listing 8.10   *A Shell Script to Parse the Output from the fetch Command*

```
#!/bin/sh

#loop across the lines that actually have values
grep '^[0-9]' | grep -v 'nan' |while read i
do
    #extract each element of the line
    for h in 'echo $i'
    do
        #if this element is data then convert it
        if echo $h | grep -q 'e'
        then
            value='echo "scale=10; $h" \
                | sed -e 's/\([0-9]\)e+*/\1^/' | bc'
            out="$out $value"
        #otherwise if it's a timestamp then reset out
        elif echo $h | grep -q '[0-9]:'
        then
            out=$h
        fi
    done
    echo $out
done
```

You can pipe the output from a fetch command straight into the shell script in Listing 8.10 and you will get back just the data that matters, converted from scientific notation into real values.

Fetching is fine if you need every data point for your intended purpose, but if you plan to do things like averaging the data, or otherwise munge it together after you fetch it, you would probably save yourself a lot of time and CPU cycles by letting RRDTool do it for you. An oft-unspoken detail about RRDTool in graph mode is that the graph element definitions are optional.

In other words, RRDTool graph doesn't necessarily have to generate graphs. It can be used from the command line to derive data for other purposes, such as sending it to external graphing tools. RRDTool performs math on internal data very quickly, so it will be much faster at doing the math for you than exporting the data to an external program.

Suppose, for example, you had a graphical widget, such as a speedometer, and you wanted it to depict the average system load for the server in Listing 8.9 for the last ten minutes. We could use RRDTool fetch, parse the data with the shell script in Listing 8.10, further parse the data to extract just avg5min, and then add the values and divide by two, but that's far more crunching than we need to do. Instead, use something similar to Listing 8.11, which is an RRDTool graph command that asks RRDTool to average the metric for us and to print it to stdout.

Listing 8.11   *Internal RRDTool Metric Averaging*

```
rrdtool graph /dev/null --start=end-600 \
'DEF:foo=server1_load.rrd:avg5min:AVERAGE'\
'PRINT:foo:AVERAGE:%1f' \
| tail -n1
```

The GD Graphics Library

In case you wondered, the bar and point charts in Figure 8.21 and Figure 8.22 were created with the GD Library. GD is an open source graphics library designed to make it easy to create images programmatically. It is implemented as a C library, but wrappers exist for Perl, Python, Ruby, and most other interpreted languages.

GD is especially good at creating instrumentation widgets, such as speedometers. Figure 8.23 is a simple gauge created with the Perl GD::Dashboard module.

Figure 8.23   A gauge created with GD::Dashboard

GD also has excellent built-in charting and graphing capabilities. As mentioned previously, the graphs in Figures 8.21 and 8.22 were created with GD; specifically, Perl's GD::Graph module. RRDs are good for dynamic data that is constantly streaming in, but GD is good in data analysis situations in which the data is static, such as one-time log analysis or creating one-off graphs for presentations. Speaking of presentations, Figure 8.24 shows off some of the whiz-bang 3D capabilities of GD using the GD::Chart Perl module. The GD library can be obtained from www.boutell.com/gd/. The various Perl modules are all available from www.cpan.org.

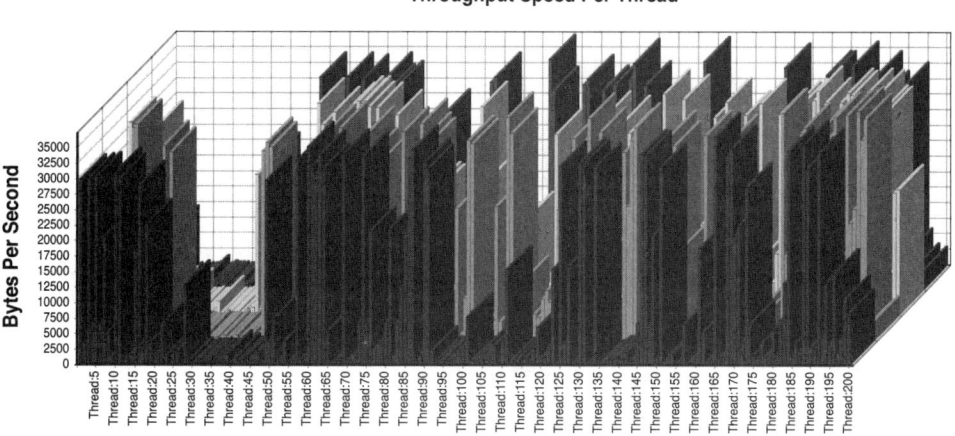

Figure 8.24   A 3D bar graph depicting execution time per thread from a web server performance test, created with GD::Chart

For a more web-friendly graph library, check out D3.js, a JavaScript library with powerful visualization components that are far more polished than their GD counterparts: http://d3js.org.

NagVis

Second to shiny little widgets, interactive maps seem to be the most requested form of management interface. These are, literally, maps with little blinking lights on them, such as the kind you see if you've ever taken a tour of a power station or water treatment plant. NagVis, a PHP tool available from www.nagvis.org, can use Nagios status data to animate status indicators on a graphic, such as a map flowchart or network diagram.

NagVis is easy to install and has a definition syntax similar to Nagios itself. Download NagiVis, untar it, and place the nagvis folder inside your Nagios share folder. There, the package should work out of the box. Figure 8.25 is a map of moderately sized corporate email infrastructure. In the NagVis map file, you define which Nagios services to map to which status indicators. When the service is okay, according to Nagios, its status indicator is green. If a service goes into a warning state, NagVis changes its status indicator appropriately, and so on. With animated GIFs, the status indicators can even be made to blink.

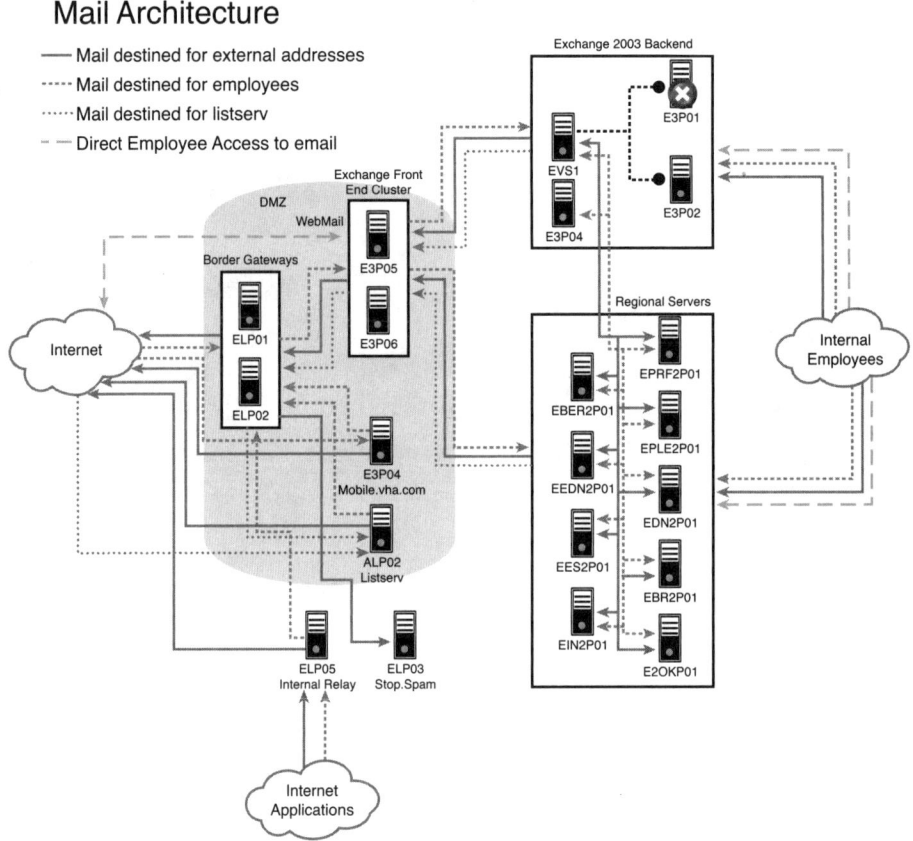

Figure 8.25 NagVis network diagram of email architecture

Each green status indicator links back to the Nagios status detail page for the service to which it refers. NagVis even provides mouseovers, so when you point at a status indicator, you get some information about the service from Nagios. In Figure 8.25, all the hosts are alive except for E3P01 in the upper-right corner. NagVis works with any static image, so you can get as creative as you want. Figure 8.26 is a map in the geography sense of the word.

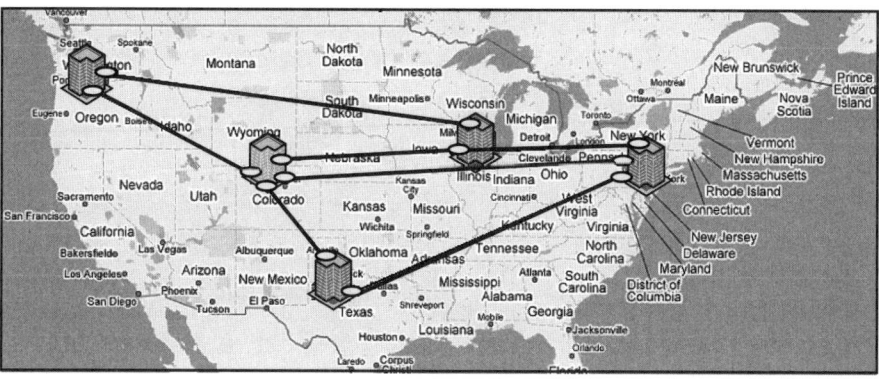

Figure 8.26   NagVis interactive map

GraphViz

GraphViz is an open source tool for programmatically creating graphs that represent struc-
tural information like networks and flow charts. It does the same sort of thing Visio does, only
programmatically instead of interactively. GraphViz is implemented as a textual language
called dot. You create a text file, similar to source code, in the dot language, and call one of
several GraphViz interpreters to compile it into an image. Each interpreter uses a different
layout algorithm, and because of this, subtle differences exist in the syntax they each support.

GraphViz diagrams are excellent when you want to show relationships between a large
number of entities. It is already a very popular tool in the sysadmin and security communities
because it works well for log analysis. Figure 8.27 is a very simple GraphViz diagram that
was generated from an NMap scan of three hosts.

GraphViz makes up for many of the shortcomings of the Nagios statusmap.cgi, but it is
also good at modeling data from log files and RRDs to spot strange behavior, or cliques. For
example, plotting a GraphViz diagram of hosts with CPU, Network, and Memory utilization
statistics causes the boxes that use more CPU than memory to cluster together in a group.

In practice, most people don't type out the dot files required to create an image. Rather,
most people use dot file creation utilities and GraphViz wrappers to do the dot file and/
or image creation. The preceding graph was created from a Perl script I wrote using the
GraphViz Perl module.

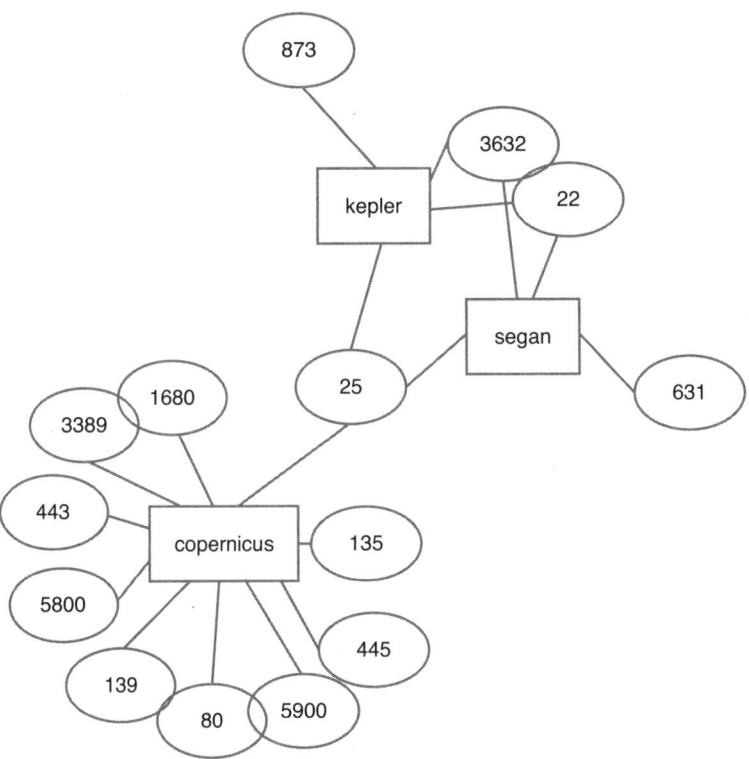

Figure 8.27   GraphViz diagram of open ports from NMap data

If writing graphing engines is not your bag, I point you to the excellent shell-based GraphViz wrapper, Afterglow, which was written by Raffael Marty. Afterglow can be used from the command line and lets you easily define colorization mappings and filters. For example, it's possible to tell Afterglow to draw only hosts that have three or more outbound connections. Afterglow's syntax allows you to quickly draw and redraw graphs by altering arguments on the command line; even if you do write code, it's worth checking out. Figure 8.28 is a GraphViz diagram created with Afterglow.

Sparklines

Sparklines are a concept introduced by Edward Tufte in his recent book, *Beautiful Evidence.*[5] Sparklines, in Tufte's words, are "intense, simple, word-sized graphics." They are drawn in one of two ways: as a miniature line graph or a miniature bar graph. Sparklines are intended by Tufte to be viewed in the context of some text, as sort of an inline footnote. Figure 8.29 shows the page hits per day of a web site for the last three months.

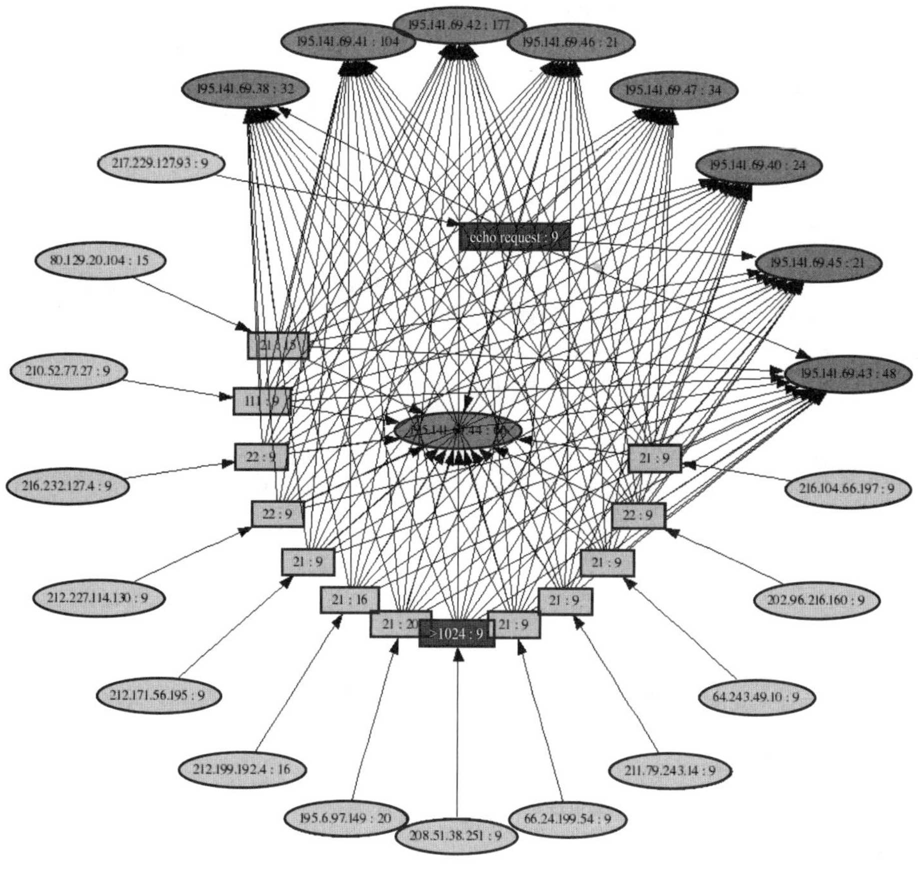

Figure 8.28   Afterglow-generated GraphViz diagram

Figure 8.29   Web page hits-per-day sparkline

The beginning and ending points are marked with a red dot and labeled. The highest and lowest points on the graph are also labeled. Sparklines are a fantastic idea for people who build monitoring dashboards of any type. I think they make a perfect replacement for "ometers," because they display much more information in the same amount of space, fit perfectly in HTML tables, and have a spiffy EKG-like[6] feel. The bar graph sparklines are equally interesting for depicting the history of Boolean states. For example, the bar graph sparkline in Figure 8.30 is the availability of the HTTP service on a web server for the past few months.

Figure 8.30   Web-service-availability sparkline

There are a few Sparklines implementations out there. The most popular is the Sparkline PHP Library, available from www.sparkline.org. Figures 8.29 and 8.30 were created with a Python script called sparkplot.py, which is available from http://agiletesting.blogspot.com/2005/04/sparkplot-creating-sparklines-with.html.

As described in Chapter 4, "Configuring Nagios," Nagios allows you to specify per-service graphics in the web interface. Sparklines are a perfect fit for this space, giving you a short history for each service inline in the user interface. Figure 8.31, for example, has a sparkline under the fully qualified domain name of the server in the status detail CGI screen of Nagios.

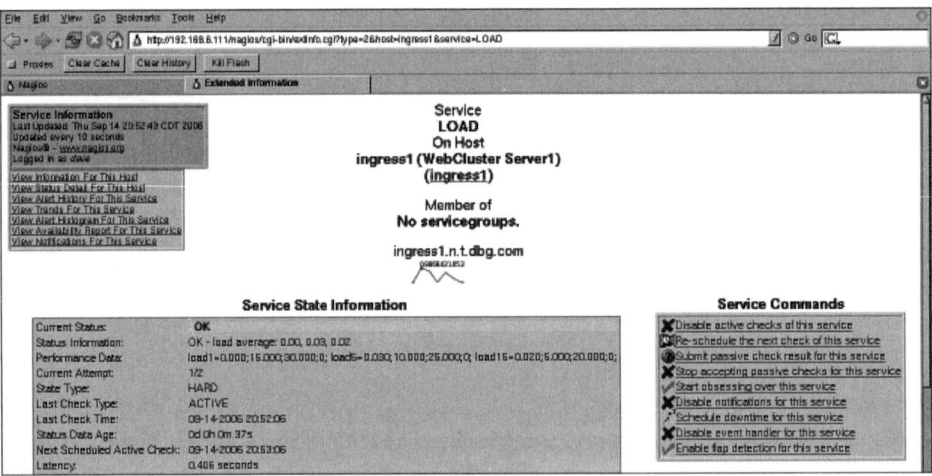

Figure 8.31   Sparkline embedded into the Nagios web interface

Force Directed Graphs with jsvis

The problem with static graphics is fitting everything in. Tools like GraphViz and LGL[7] use complex layout algorithms that maximize for the most efficient use of space while maintaining readability, but large graphs can still become cluttered, limiting their usefulness. If you've ever wished you could interact with a network graph in real-time, jsvis might be for you.

Written by Kyle Scholz, jsvis is a freely available JavaScript applet that implements force-directed graphs in a web interface. If you've ever used the visual thesaurus (www.visualthesaurus.com/), you are familiar with the concept. Force-directed graphs are very much like GraphViz diagrams that you can interact with. If the graph contains so many servers that they are obscuring each other, for example, you can move the nodes you want to examine into view.

I think force-directed graphs have huge potential in a systems monitoring context. Far less complex and more organic than 3D visualization, they are the best way I've seen to model a large number of network nodes in a small space. Kyle's blog contains the code, interactive samples, and even a tutorial to get you started coding. I highly recommend a visit: www.kylescholz.com/blog/2006/06/force_directed_graphs_in_javas.html.

Animation is certainly the next step in data visualization for monitoring systems and intrusion detection and prevention. The integration of NetFlow[8] data into routers and network gear has accelerated the use of animated visualization tools[9] and, as the trend toward the animated display of monitoring data continues, I expect to see many more solutions like jsvis being integrated into web interfaces. This is undoubtedly a good thing for humans who struggle for visibility in complex systems. If you are building custom management interfaces, there's no reason you couldn't get ahead of the power curve today with jsvis.

End Notes

[1] Yes, in reference to the television program: *The X-Files*.

[2] To paraphrase Kenny Rodgers, there'll be time enough for mucking when the data's stored.

[3] Ideally, pick one with the polling code and graphing code in separate files.

[4] Well, most of them.

[5] The sparklines chapter is available online: www.edwardtufte.com/bboard/q-and-a-fetch-msg?msg_id=0001OR&topic_id=1.

[6] Electrocardiogram; see http://en.wikipedia.org/wiki/Electrocardiogram.

[7] http://apropos.icmb.utexas.edu/lgl/

[8] A Cisco standard to depict network traffic flows as a unidirectional sequence of packets all sharing the same source and destination IP address, source and destination port, and IP protocol. Current open source NetFlow analysis tools exist to visualize these network flows in real-time or as historical data in a forensics context.

[9] The gpl cube of potential doom: www.kismetwireless.net/doomcube/, NvisionIP: http://security.ncsa.uiuc.edu/distribution/NVisionIPDownLoad.html, xovi: www.doxpara.com/?q=node/1133.

## CHAPTER 9

Nagios XI

In 2009, Nagios Enterprises, the corporation formed by Nagios creator Ethan Galstad, launched Nagios XI, a commercial version of Nagios. XI truly is an amazing accomplishment. You need to know next to nothing to use it, and yet the first eight chapters of this book are prerequisite to your understanding it. But now that you have a good handle on how Nagios and the various add-ons surrounding it work, we can finally examine XI and see if it might be a good fit for you.

What Is It?

After the release of 3.0, Nagios was, it seemed, in danger of becoming a victim of its own success. Sysadmins who knew and loved it were happy to see it continue in the way it always had, but its popularity had risen to the point that a different and more populous group of potential end users had taken notice, and with them, Nagios wasn't comparing favorably with newer, prettier, and less flexible commercial competitors.

This new breed of user was quite vocal and had a few very specific gripes. First they found Nagios's configuration syntax unwieldy, to say nothing of the intolerable notion of (gasp) editing text files by hand. Second, they found the Nagios web interface, with its C-based CGI and lack of integrated time-series data, unforgivably old-fashioned. Finally they had no idea what to make of the fact that there was no database back-end. Jiminy Christmas—wrist watches and garbage disposals run MySQL these days! How was one to take seriously a monitoring system that didn't?

For this considerable subset of users, Nagios's price tag didn't make up for its abhorrent lack of bling, and answers to the effect that all these things could be rectified with add-ons fell on deaf Bluetooth earpieces. Add-on options were birds in the bush, and they would rather pay for a bird in the hand than go beating around the bush themselves for free.

XI might best be called the perfect compromise between maintaining the power and flexibility of Nagios and providing a turnkey monitoring system that more than satiates the desires of the PHP proletariat. But that description sells it short; XI is much more than just a shiny interface; it represents a huge amount of custom development and integration work. Further, there is real functionality in XI that simply can't be found in Nagios Core. But neither can it be called a new monitoring system in its own right, because in very important ways, it remains Nagios and retains all the flexibility and power that I've described in the previous chapters. Everything I've written up to now about the underlying architecture, plug-ins, scalability, and even advanced visualization, is applicable to Nagios XI.

It'll be easier to just show you.

How Does It Work?

Figure 9.1 is a rough sketch of the Nagios XI architecture. As you can see, all host and service monitoring, as well as notification, escalation, and so on, relies on an unmodified Nagios Core daemon, so any preexisting plug-ins or customization you might have can be made to work under XI. The NDOUtils plug-in (described in Chapter 7, "Scaling Nagios") has been enabled and configured to replicate state information from Nagios Core into a MySQL database. Here is the primary information hand-off between Core and XI; Nagios XI reads this database to glean information about the current state of hosts and services, as well as the Core daemon itself. This adds an information layer to Core that can be consumed by third-party UIs as well as your own custom integration scripts.

The NagiosQL add-on (described in Chapter 5, "Bootstrapping the Nagios Config Files") provides the hooks necessary to modify the Nagios Core configuration from the XI interface. Every parameter that can be configured in the flat files may be set via the web interface using the customized NagiosQL forms in the "Advanced Configuration" section of the XI interface. Although these forms are well integrated into XI, and retain an XI look and feel, there is a bit of a line in the sand between NagiosQL-driven core configuration, which is referred to as "advanced" in the XI interface, and the configuration parameters that are specific to XI itself.

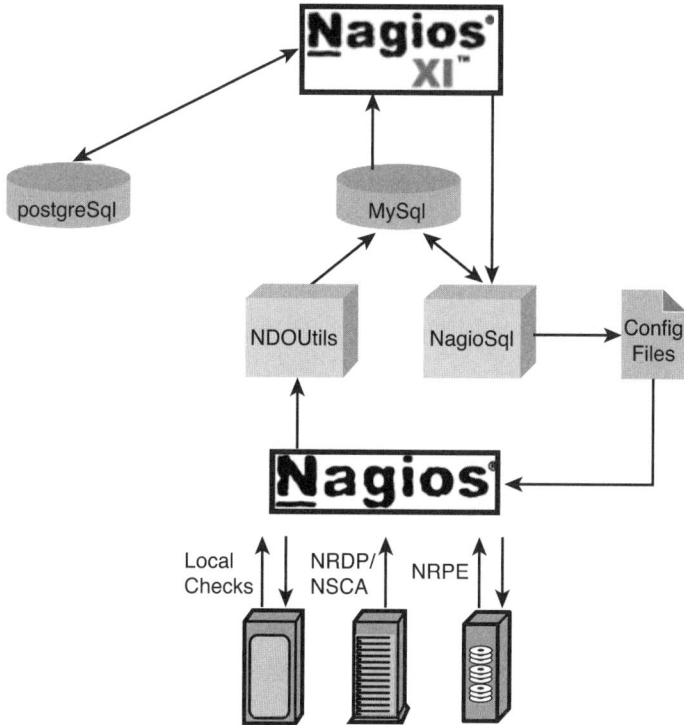

Figure 9.1 The Nagios XI architecture, simplified

XI goes beyond presenting a simple web wrapper to the Nagios Core configuration files, providing in addition a litany of semiautomated wizards and autodiscovery tools to ease the burden of initial and ongoing host and service configuration. I talk more about these later, but suffice to say that it is the intention of the XI creators to isolate the majority of XI users from the intricacies of the Nagios core configuration to the extent that they never need to know what a check command is, much less a template. This makes it possible for monitoring configuration, traditionally an operations task, to be delegated to first-level support types, or in some environments, even to normal users. More clueful administrators who need to customize this or that can still do so, without editing the config files by hand, using the NagiosQL-driven advanced configuration tool.

Configuration created by NagiosQL is automatically written to text configuration files in etc/nagios and is read by the Core daemon from these flat files in the usual fashion. Although it's technically possible to hand edit these configuration files, you will gain nothing because NagiosQL will eventually overwrite any changes you make. If you have your own configuration-generating automation (like Check_MK), or preexisting configuration that you do not want to import into NagiosQL, or even if you're a curmudgeon who just prefers to manually edit the configuration, you can still maintain static config files in etc/nagios/static, and your files will still be parsed by the Core daemon while being left alone by NagiosQL. That runs both ways; statically configured hosts and services can't be modified via the UI unless you manually import them into NagiosQL (at which point they cease to be static).

Finally, Nagios XI maintains its own Postgresql database to store various configuration parameters such as user-settings, custom dashboards, authentication info, and the like. Given the shiny new PHP interface, the simplified configuration options, and the open database back-end, Nagios XI should satisfy the complaints I'm used to hearing from corporate administrators who are in the market for a "grown up" commercial monitoring product; however, there's a lot more functionality than what I've encompassed in the architecture diagram.

What's in It for Me?

Now that we've taken a quick look at what XI is and how it works, let's take a look at how XI compares to Nagios Core and the various commercial monitoring systems with which it was designed to compete.

One Slick Interface

Given the general quality of the alternative PHP interfaces we find in the Nagios Exchange repository, the XI interface is shockingly excellent. It is certainly not yet another effort to bring the CGI interface "up to date" by replacing it with a PHP version of itself. The XI user interface is a complete rethinking of the UI, which truly takes advantage of the strengths of a web programming platform like PHP at every opportunity. Elements within dashboards can be unlocked, moved around, or even deleted to suit the preferences of the user. AJAX is employed, both to update individual information elements and to provide feedback, so that when I send a command via the UI to reschedule a service check or acknowledge an alert, a box momentarily appears to let me know my command has been accepted. One of my least favorite things about the Core UI is the way it dumps me to an acknowledgment page after I've issued a command, forcing me to manually navigate back to somewhere useful.

The traditional Nagios tables like "service detail" and "hostgroup grid" still exist, but are implemented as repurposable widgets that I can use to build custom dashboards. New tables have been added, a few of which are very dense and handy, like the "minemap" visualization pictured in Figure 9.2.

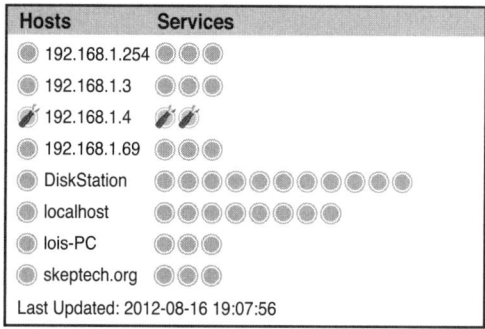

Figure 9.2 Nagios XI Minemap

One of my favorite Nagios Core views is the hostgroup grid, which shows at a glance the state of entire hostgroups, including their services. This is one of the more dense status visualizations available in the old UI; unfortunately, I still need to scroll around to see everything in my environment. The Minemap visualization, by comparison, shows the same information in a much smaller amount of screen space, enabling me to get a coherent, uncluttered, detailed service-level visualization of my entire network on a single screen.

Integrated Time Series Data

PNP4Nagios (described in Chapter 8, "Visualization") is integrated out of the box, and definitions exist for all the included plug-ins. This means that without any additional configuration whatsoever you get time series data for every service you configure. The RRDTool graphs are so well integrated into the UI that the uninitiated user would never guess PNP or RRDTool were community-sourced add-ons, so you get a snazzy UI without losing any of the power and flexibility that these community-driven development efforts provide.

In addition to the RRDTool graphs, small bar-graph visualizations for metrics collected by the Nagios Core daemon, as well as remote execution tools like NRPE, are sprinkled throughout the interface. These do a great job of conveying capacity planning info at a glance, as well as giving the UI a very polished look.

Rounding out the time series visualization is a Graph Explorer tool, which allows you to draw, among other things, ad hoc time series and stacked time series graphs. The graph explorer uses a commercial JavaScript library from HiCharts.com and looks quite elegant. The data comes from the RRD's resident on the Nagios server via rrdtool fetch and is provided to the end-user's browser to compute the graph locally. This saves the server's

CPU and provides a snappy, feature-rich data visualization, allowing you to scale the graph by dragging to select a range and providing pop-up numerical values when you mouse over any data areas. The stacked time series graphs include time-shifted historical data, so you can easily compare today's data to that of yesterday, and so on.

Modularized Components

The UI as a whole is highly modular, incorporating add-on components to implement extra features. This enables the XI developers to quickly react to the needs of the user community by adding features to the UI as needed or even adding custom developing features for larger end users with special needs. A notable example is the Operations screen depicted in Figure 9.3, which is intended to be displayed on a dedicated screen in a Network Operations Center. In addition to this and other single-page summaries, custom views can be configured to rotate between pages with more detailed information on timed intervals. I bring up these little summary views because seeing them so prominently displayed in the XI interface hits home for both the extent to which the Nagios developers are listening to the needs of the community and their eagerness to satisfy those needs now that incremental progress in the UI is possible.

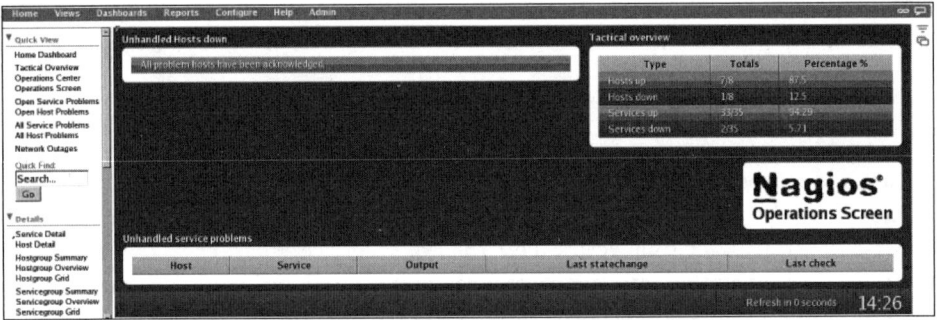

Figure 9.3   Nagios XI Operations screen

Finally! Acknowledgments and Scheduled Downtime for Multiple Hostsv

Another component that implements a feature for which the core community has been begging for years is the Mass Acknowledgment Component. This allows an admin to schedule downtime and acknowledge problems for groups of hosts and services. I know more than one sysadmin who would purchase XI for this feature alone.

Enhanced Reporting and Advanced Visualization

The XI developers are not solely focused on the community, however, as a quick glance at the Reporting tab in XI shows; they are proactively exploring some interesting data visualization techniques from the neoformix data-visualization field. Components that implement heat maps, force directed graphs, and stream graphs, as depicted in Figure 9.4, have been added to the classic reporting options. Several shiny new implementations of the core reports are also provided, each of which I find generally cleaner than their legacy counterparts and more likely to impress the wearers of neckties and high heels in our lives. The new reports may be exported in CSV and PDF formats with the click of a button. The button, which links to a predictable URL, makes it possible for the shorts and t-shirt wearers among us to automatically grab the reports with tools like curl and wget.

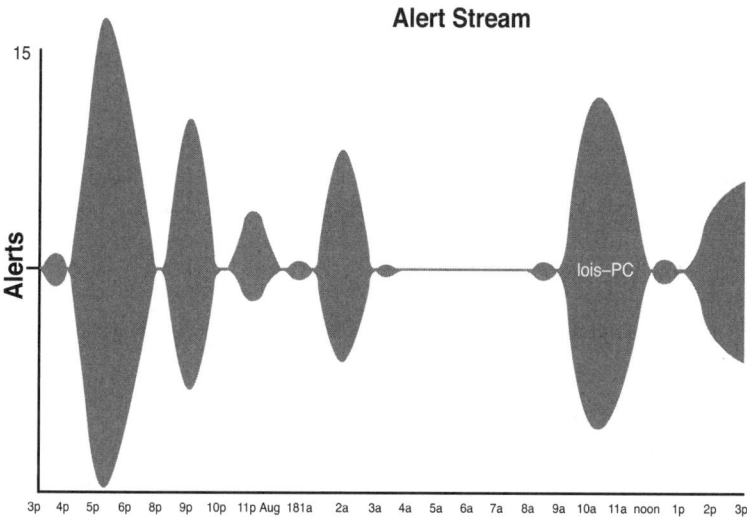

From: 2012-08-17 16:46:56 To: 2012-08-18 16:46:56

The alert stream provides a visual representation of host and service alerts over time.

Clicking on a host name will cause the graph to drill down to show service alerts for that particular host.

Figure 9.4 Nagios XI Stream Graph component

Nagvis

Nagvis, (described in Chapter 8) is installed and available in the Maps section of the Home view. Setting up your own NagVis diagrams couldn't be easier. First, copy your map or diagram graphic to /usr/local/nagvis/share/userfiles/images/maps, launch the Nagvis tool in the XI UI, select Manage Maps from the options menu, and create a new map, pointing the Background to the map you uploaded. Finally, open your map using the Open menu, and add status icons to it by selecting Add Icon from the Map menu.

Business Processes

Nagios XI contains wrapper logic for grouping individual services into higher-level entities called business processes. The intent here is to implement what the Gardiner Group calls Business Application Monitoring, or BAM. BAM attempts to provide real-time status for critical business entities like a sales catalog web site or corporate email. Nagios XI implements BAM by breaking a high-level concept like "corporate email," into its requisite pieces, such as Mail Transfer Agents, Mail Exchangers, Groupware systems, and Databases, and then quantifying the relative importance of each of the services that make up those pieces as well as describing dependency relationships between them.

XI Business Process groups contain services that are said to be "essential" or "non-essential." A database service in our example might be considered essential, whereas the SMTP port on a single mail exchanger might be "non-essential" (because they are usually redundant, and even if they go down, the mail will queue somewhere else). When any essential service or the combination of all non-essential services goes critical, the XI business process logic registers this as a "problem."

Each business process group contains critical and warning thresholds that depend on the number of problems that are occurring in the group. In our example, we might imagine two business process groups, one for SMTP speakers (MXs and MTAs) and one for SQL-speakers (groupware systems and DBs). If the latter group registers a single problem because a database is down, that might throw the whole group into a warning state.

Business process groups can contain other nested business process groups, and so on. Our top-level entity, corporate email, is therefore just a business process group that contains the two groups previously described. It is configured like the other two groups so that a single "problem" in any of the nested groups causes it to go into a warning state. Finally, notification commands can be assigned on each business process group in the same way they are assigned to individual host and service events. Additionally, visualization widgets exist for the top-level groups. These can be added to any dashboard or view, and they allow the user to drill down into the groups to see what services or subgroups constitute them.

Integrated Plug-ins and Configuration Wizards

The core installation of Nagios XI includes all the plug-ins in the standard plug-ins package, as well as NRPE, NSCA, and NRDP. In addition to all the plug-ins being preinstalled, the XI developers have provided a plethora of semiautomated configuration wizards, which, given the bare-minimum information about a host, take care of the initial setup as well as adding and modifying services on already configured hosts.

If you consult the official XI documentation at

`http://library.nagios.com/library/products/nagiosxi/documentation,`

you'll quickly discover that the wizards are the preferred method for host and service configuration. With names like Exchange Server, website, and Windows Workstation, they make setting up new hosts and services easy enough that these tasks can be delegated to first-level support techs, or even end users. The autodiscovery wizard is capable of bootstrapping an environment given only a CIDR netblock to start with, and it does a good job of initial setup. To add NRPE-based host checks or other services after the fact, run the appropriate wizard on the preexisting host.

For example, if Server1 was created with the autodiscovery wizard, and you now want to add NRPE checks to get CPU, memory, and disk information from the host, you must first install NRPE on Server1. If Server1 doesn't already have NRPE on it, and is one of several common server types, such as a Windows 200X server, Red Hat, or Ubuntu, the XI developers have an agent package designed to work with XI specifically at:

`http://assets.nagios.com/downloads/nagiosxi/wizards`

After the agent is installed on Server1, run the NRPE Wizard on the server from the configuration tab of the XI user interface, as shown in Figure 9.5, entering the IP or FQDN of the server, and choosing the type from the drop-down list. The wizard will then display a preconfigured subset of available check commands relevant to your server type, and provide text-entry fields for you to specify custom settings or additional commands if you'd like.

As I said earlier, static configuration files may still be maintained in etc/nagios/static. So it's entirely possible to run your own scripts, or autogeneration tools like those included with check_mk, provided you configure them to write their configuration to the static directory. I can't deny that the automated configuration features in XI have, perhaps ironically, complicated things a bit for those of us who have reason to maintain the configuration manually. In the Nagios Core universe, there is a single way to configure Nagios (text files). However, there are three ways to configure Nagios Core in the XI universe (text files, NagiosQL, and XI Wizards), and although the three coexist well enough, it can become burdensome to ensure a uniformity of parameters if the administrators mix and match their configuration methodologies in XI. I'll give you an example.

Larry, his brother Darryl, and his other brother Darryl all work at Bloody Stump Lumber Mill, where they recently purchased a Nagios XI server to monitor their growing sales web-application server farm. Larry was a UNIX admin in college, so he prefers to edit the config files. Darryl likes to have fine-grained control over the config, but isn't very good in vim, so he uses the XI advanced configuration section, and other Darryl would rather be watching football, so he just runs the wizard for everything. Each of the brothers has a server running sshd that he wants to configure in XI.

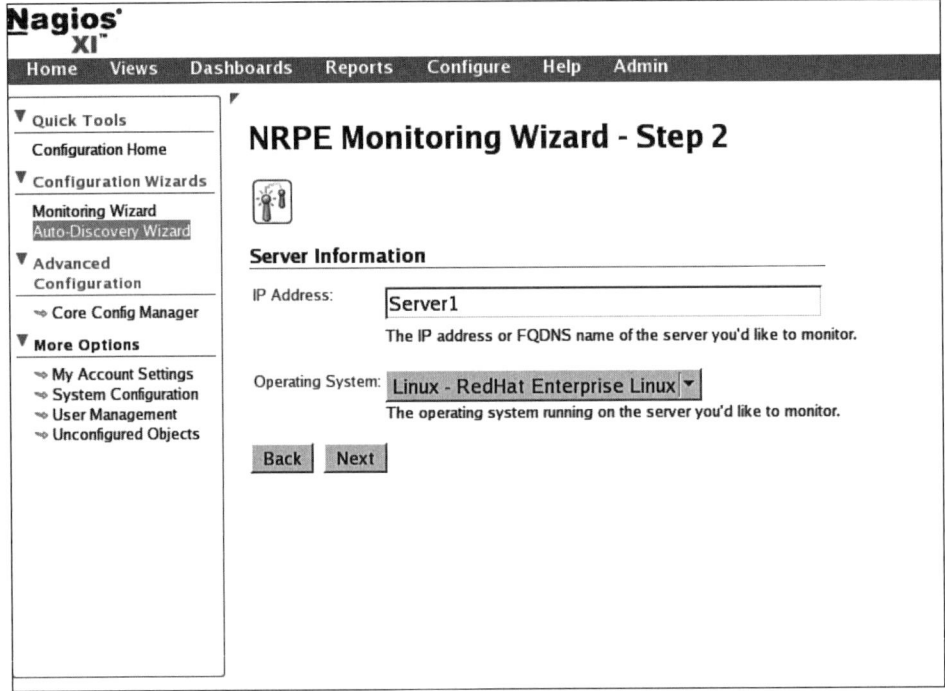

Figure 9.5   The Nagios XI NRPE Wizard

When other Darryl runs the Autodiscovery Wizard on his server's IP, XI scans the host and automatically configures a host check and a check_tcp service check for the SSH port. It then pushes the config to NagiosQL, which commits it to the DB, writes out the configuration, and restarts the daemon.

Darryl meanwhile, sets up his host using the NagiosQL forms directly, but instead of choosing check_tcp, he chooses the check_ssh service, which does pretty much the same thing, but returns slightly different output. He also names the service "ssh" instead of "SSH" like the wizard does.

Larry, meanwhile, has really done his homework. He already has a servicegroup for ssh servers in the static config files he created, so rather than doing all the typing and clicking that his brothers do, he simply adds his server to the ssh_servers servicegroup, and the rest takes care of itself. The problem is, his servicegroup inherits a different set of templates than NagiosQL, so although his service check uses the same name and check command as the wizard, his polling interval is different, and he has a different notification target for service warnings.

In this way, the brothers end up with three different definitions for the same service, which might not be a problem immediately, but will cause all manner of headaches if and when they want to integrate Nagios with another tool, or generally try to do any sort of automation using their monitoring server.

I admit these sorts of disconnects are possible with text configuration files, but my point is the text configuration encourages administrators to use templates to normalize the configuration, like Larry did in the previous example. The automated tools by comparison encourage isolating the configuration at the host level, because it's easier for the automated tools to parse them that way. Thus, in Larry's configuration, we find a single services.cfg wherein every service is defined and assigned a hostgroup, whereas in NagiosQL's configuration, we find a services directory with a single file for each host. The former makes it pretty easy to verify that all the service checks for every host are implemented in the same way. The later makes it much more difficult.

Further, in my experience, the disdain that people like Larry naturally feel for people like other Darryl generally discourages them from paying close attention to what people like other Darryl are doing. In fact, merely inviting other Darryl to configure the monitoring server with wizards might trigger a tendency in Larry to go off on his own and "do it the right way" using well-written static config files, which only exacerbates the problem by more widely diverging the configuration paths.

Whether this will be a problem in your shop will depend on how many hands are stirring the pot and the extent to which the more clueful users are aware of the potential problem. The idea of delegating the configs is certainly tempting, and I'm not saying you shouldn't. If you do, my advice would be to use either the wizards or static config for service and host *creation*, and avoid using NagiosQL directly if you can avoid it (you could still safely use it for host and service *modification*). That way, you can carefully set up the static config to ensure that it references the wizard templates, or simply copy definitions from the NagiosQL files, and everything should remain pretty much uniform.

Automated Configuration for Passive Checks

Another very cool bit of functionality that is related to automated configuration in Nagios XI is the Unconfigured Objects feature. In the event that XI receives a passive check result for a host or service that it doesn't know about, it automatically generates an inert configuration for that host or service and places it in the Unconfigured Objects section of the Configure tab.

Administrators may then approve the inert objects, and they will become part of the running configuration. Good stuff.

Operational Improvements

In addition to the myriad functional improvements in Nagios XI, several maintenance-related features exist that make it easier to manage the Nagios server itself.

Backups

Out of the box, XI takes a snapshot of the running configuration each time it changes. These configuration snapshots can be downloaded from the UI in an automated fashion using tools like curl or wget. It can be used to restore the configuration in the event the monitoring system kicks the bucket, or it can roll it back to a prior version if someone made an inappropriate change. A real system backup, including historical state and metric data, involves a lot more than just the configuration files, however. Remember, XI maintains three databases and has untold amounts of performance data stored in RRDs, not to mention the Nagios Core state file and logs. For detailed instructions on properly backing up your XI install, see:

```
http://assets.nagios.com/downloads/nagiosxi/docs/Backing_Up_And_
Restoring_XI.pdf
```

User Management

Account management is more important in XI, especially when individual users are encouraged to change configuration parameters and create new hosts and services. Individual users in XI also have the ability to configure the interface with custom views and dashboards as they see fit. For these reasons, XI must track users in its own database rather than leaving it up to Apache to sort out like the Nagios Core UI does. Account management is well done in XI and generally behaves in a manner that enterprise users expect. Access control exists to prevent individual accounts from making modifications, and components exist to enable XI to use LDAP servers. Nagios has published official documentation on multitenant setups, where, for example, access to a Nagios server hosted by a service provider is shared by multiple customers. This documentation resides at:

```
http://assets.nagios.com/downloads/nagiosxi/docs/XI_Multi-Tenancy.pdf
```

Daemon Status

As depicted in Figure 9.6, the XI interface provides an array of detailed of information about the Core daemon process. This includes metric values for the server hardware as well as performance metrics internal to the daemon itself. A real-time graph of the event queue displays reaper and service check events scheduled 5 minutes into the future. This really is fantastic capacity planning info of a quality I've never seen in any monitoring system.

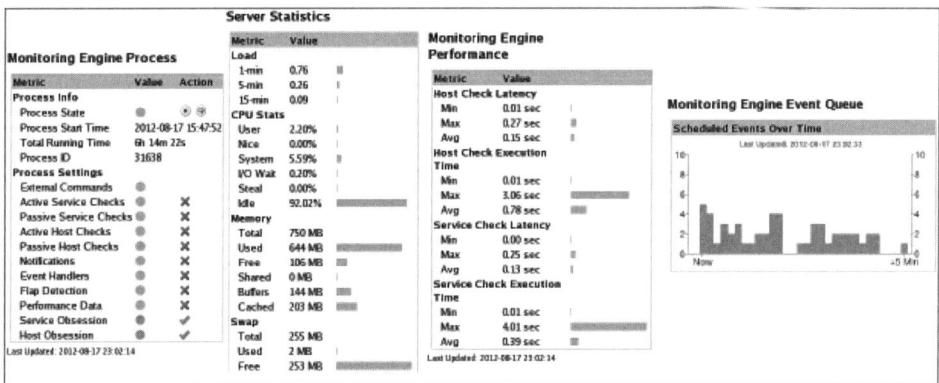

Figure 9.6   Detailed daemon statistics

How Do I Get My Hands on It?

Fully functional demo versions of XI (60-day expiration) are available from nagios.org. You may download self-contained installers, or VMware disk images with XI preinstalled. The latter can be run by any system that supports the free vmplayer utility, while the former requires a relatively recent Red Hat or CentOS install.

The reason for the RHEL dependency is the dizzying array of packages that must exist for XI to run. The XI developers have chosen to rely heavily on YUM to satisfy the requisite dependencies, so although it's possible to run XI on other distros, you won't be able to use the official installation script to get it up and running on anything other than a Red Hat or CentOS system.

## CHAPTER 10

The Nagios Event Broker Interface

In this chapter, we're going to delve into the inner workings of Nagios by exploring the Nagios Event Broker. The Event Broker is new to 2.0, and it is, bar none, the most powerful interface available to Nagios; however, actually wielding it requires some modest knowledge of C programming. Don't let that scare you, however; if you possess even a passing familiarity with C, the information I present in this chapter should have you well on your way to extending Nagios's functionality to your heart's content.

Function References and Callbacks in C

If C programming isn't something you do often, you may not have ever used function pointers. If you are adept at C, feel free to skip this section. Function pointers are equivalent to variable pointers, except they point to a memory address that corresponds to a function instead of a variable of some type. If you understand pointers, they work the exact same way and their syntax is what you would expect, but they are rarely covered in C programming books. I think this is because it's hard to come up with simple examples in which they might be useful. This is a shame, because they enable some elegant software engineering in larger C programs, such as Nagios.

Nagios uses function pointers often to implement callbacks. Callbacks are functions that take pointers to other functions as initialization arguments. When interesting events occur, Nagios can use the passed function pointers to call back to event handlers that are interested in that particular type of event. But before we get into all of that, take a look at Listing 10.1, which outlines the use of a function pointer.

Listing 10.1   *Using a Function Pointer*

```
void main(){

/* ******************************************
    Here we have two functions, one that converts Celsius
        to Fahrenheit and one that does the opposite.
        ****************************************** */

    int c2f( int c ) { return (9/5)*c+32; }
    int f2c( int f ) { return (5/9)*(f-32);}


/* ******************************************
    The convert function acts as an interface to
    the actual math functions. It takes a function
    pointer as one of its init arguments.
        ****************************************** */

    int convert(int input, int (*fPointer)(int)){
        return fPointer(input) ;

}

/* ******************************************
    Now we can call the convert function
    whenever we want, and we can tell it which
    conversion to do by passing it a pointer to
    either f2c or c2f, like below.
        ****************************************** */
void Go(){
        int result=convert('72',&f2c);

}
}
```

So you can see why nobody would teach you this in a programming book. Using normal conditional logic, such as a switch or if/else loop, would be a much more straightforward way to choose between f2c and c2f. In this example, and probably any other simple example you could think of, there are better ways to do things than using function pointers. Function pointers begin to shine, however, when things get a bit more complicated. The primary magic you need to understand, in order to write NEB modules, is the convert function declaration line:

```
int convert(int input, int (*fPointer)(int)){
```

The first argument is a normal integer called input, but the second argument is strange indeed: int (*fPointer)(int). If you try to parse it as a variable argument, it doesn't make any sense, but this isn't a variable argument; it's actually a minifunction declaration inside the convert declaration. So, in plain English, the convert function takes two arguments. The first is an `int` called input, and the second is a pointer called `fPointer`, which points to another function that takes a single int as an argument and returns an int. If the `fPointer` function took an `int` and a `float`, the declaration would have looked like this:

```
int convert(int input, int (*fPointer)(int,float)){
```

The convert function just turns around and calls whichever conversion function to which it is passed a pointer. Passing a function pointer to convert is just like passing any other pointer: We use the '&' operator to pass the memory address that corresponds to the conversion function we want to use. In Listing 10.1, we chose to do an `f2c` conversion, so we passed &f2c ('the address of f2c') to convert. That was a crash course in function pointers. If you still don't grok, no worries; I recommend checking out www.newty.de/fpt/intro.html.

The NEB Architecture

As depicted in Figure 10.1, the Event Broker itself is a software layer between Nagios and the NEB modules. Nagios notifies the Event Broker of interesting events. The Event Broker's job is to figure out which modules, if any, are interested in the events and to create and pass out memory handles to the modules, which the modules can use to get work done.

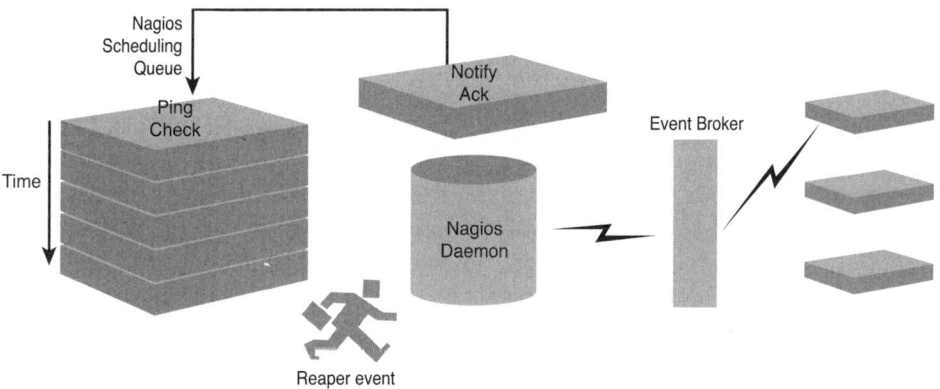

Figure 10.1 The NEB architecture from the perspective of the Event Broker

NEB modules are shared libraries written in either C or C++. The NEB module registers for the types of events it is interested in and provides function pointers to functions that presumably do things with the events they receive. Each NEB module is required to have an entry and exit function and, beyond that, can do pretty much anything it wants. The interesting thing about this architecture is that Nagios globally scopes just about everything,[1] so from the perspective of the NEB module, the architecture looks more like the one depicted in Figure 10.2.

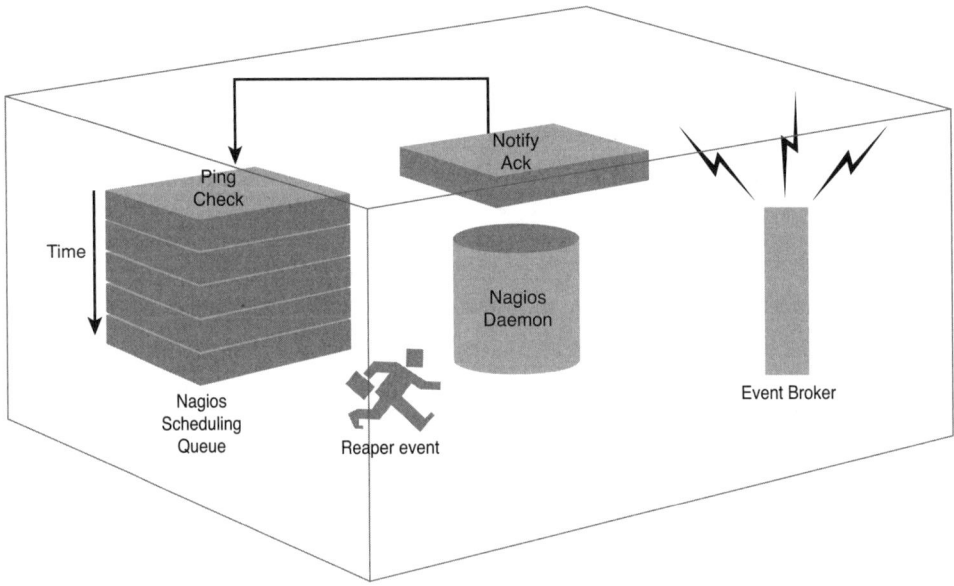

Figure 10.2   The NEB architecture from the perspective of an NEB module

That is to say, because pretty much all the interesting functions and structs are globally scoped, as long as Nagios's execution pointer is in the module's address space, the module has the power to change anything it wants to change about the entire runtime environment. It can insert and remove events from the scheduling queue, and it can turn on or off notifications. In a nutshell, anything that can be changed at runtime can be changed by the module. You might think that the module would have a limited opportunity to do these things because Nagios only runs its callbacks when interesting events to which it is subscribed happen, but because the functions to insert events in the queue are globally available, the module could, conceivably, schedule its own callback routines in a timed fashion when it is first initialized.

So what sorts of events can a module subscribe to? In Nagios 3.2.0, the version of Nagios I'm using as I write this, there are 31 total callback types, although some of them are reserved for future use. These constants are defined in nebcallbacks.h, in the includes directory of the tarball. Table 10.1 lists the callback type constants.

Table 10.1 *NEB Callback Types*

NEBCALLBACK_RESERVED0	NEBCALLBACK_DOWNTIME_DATA
NEBCALLBACK_RESERVED1	NEBCALLBACK_FLAPPING_DATA
NEBCALLBACK_RESERVED3	NEBCALLBACK_PROGRAM_STATUS_DATA
NEBCALLBACK_RESERVED4	NEBCALLBACK_HOST_STATUS_DATA
NEBCALLBACK_RAW_DATA	NEBCALLBACK_SERVICE_STATUS_DATA
NEBCALLBACK_NEB_DATA	NEBCALLBACK_ADAPTIVE_PROGRAM_DATA
NEBCALLBACK_PROCESS_DATA	NEBCALLBACK_ADAPTIVE_HOST_DATA
NEBCALLBACK_TIMED_EVENT_DATA	NEBCALLBACK_ADAPTIVE_SERVICE_DATA
NEBCALLBACK_LOG_DATA	NEBCALLBACK_EXTERNAL_COMMAND_DATA
NEBCALLBACK_SYSTEM_COMMAND_DATA	NEBCALLBACK_AGGREGATED_STATUS_DATA
NEBCALLBACK_EVENT_HANDLER_DATA	NEBCALLBACK_RETENTION_DATA
NEBCALLBACK_NOTIFICATION_DATA	NEBCALLBACK_CONTACT_NOTIFICATION_DATA
NEBCALLBACK_SERVICE_CHECK_DATA	NEBCALLBACK_CONTACT_NOTIFICATION_METHOD_DATA
NEBCALLBACK_HOST_CHECK_DATA	NEBCALLBACK_ACKNOWLEDGEMENT_DATA
NEBCALLBACK_COMMENT_DATA	NEBCALLBACK_STATE_CHANGE_DATA
NEBCALLBACK_CONTACT_STATUS_DATA	NEBCALLBACK_ADAPTIVE_CONTACT_DATA

These callback types cover every type of event that can happen in Nagios. An NEB module may register to receive information about any or all of these event types. After it initializes all the modules, the Event Broker waits for events matching the type subscribed to by the module and, upon receiving one, gives the module information about the event, as well as a handle to the relevant data structures.

For example, if the module registered for EXTERNAL_COMMAND_DATA, the Event Broker would notify it every time an external command was inserted into the command file. A handle to a struct that defined the command would accompany the notification. The module could inspect and optionally change any of the information in the command struct, or even delete it altogether. But enough talk about the architecture; the best way to learn about the NEB is to see how these modules work in practice.

Implementing a Filesystem Interface Using NEB

In this section, I'll walk through a simple NEB module I wrote, which implements a filesystem status interface to Nagios. The basic idea behind the filesystem interface is to make it easy to write shell scripts to do things such as check the status of a particular service on a certain host or output a list of all the services that are not okay. Wouldn't it be nice if the shell script could grep across a filesystem instead of parsing logs or Web pages? If we were able to tell Nagios to write a file for each service containing the service's current status code, we could do something such as the following:

```
cat /var/lib/nagios/status/host24/ssh
```

And if we got back a 0, we would know that the SSH service on host24 was okay. This also makes it trivial to get summaries of the entire environment with commands such as the following:

```
grep -rl 2 /var/lib/nagios/status/
```

This command would give us a list of all the services that were in a 2 (critical) state. This is exactly the kind of problem that NEBs were designed for.[2] First, we want Nagios to notify us of service status events. When we get them, we want to write them out to files. The module in Listing 10.2 borrows heavily from a blog post by Taylor Dondich from Groundwork, which no longer appears to be online. It is a fully functional NEB Module that implements the filesystem interface describe previously.

Listing 10.2   *An NEB Module That Implements a Filesystem Interface*

```
#ifndef NSCORE
#define NSCORE
#endif

/* include the needed Event Broker header files */
#include "../include/nebmodules.h"
#include "../include/nebcallbacks.h"
#include "../include/nebstructs.h"
#include "../include/broker.h"

/* include some Nagios stuff as well */
#include "../include/config.h"
#include "../include/common.h"
#include "../include/nagios.h"
#include "../include/objects.h"
```

```c
/*declare the handler function to make this example easier to
write about*/
int handle_service_status(int , nebstruct_service_status_data *);

// specify Event Broker API version (required)
NEB_API_VERSION(CURRENT_NEB_API_VERSION);

// our module handle
void *basic_module_handle=NULL;

/* this function gets called when the module gets loaded by
the Event Broker*/
int nebmodule_init(int flags, char *args, nebmodule *handle) {

    basic_module_handle = handle;

write_to_logs_and_console("Loading FS Module...",
                          NSLOG_INFO_MESSAGE,TRUE);

neb_register_callback(NEBCALLBACK_SERVICE_STATUS_DATA, handle,
                   '1', (void *)&handle_service_status);

    write_to_logs_and_console("Done",NSLOG_INFO_MESSAGE,TRUE);

    return 0;
}

// this is our unloading function, which gets called by the neb
int nebmodule_deinit(int flags, int reason){

write_to_logs_and_console("Unloading FS Module...",
                          NSLOG_INFO_MESSAGE,TRUE);

    return 0;
}

/*this function handles service status updates from the
Event Broker*/
int handle_service_status(int neb_event_type,
                          nebstruct_service_status_data *ds){

    //get a handle to the service struct
    service *svc = ds->object_ptr;

    //create some name buffers for various output
    char outbuf[100];
    char host_path[100];
    char service_path[10 0];

    //create a logging string and write it out
    sprintf( outbuf,"Caught status code: '%i' from host '%s' for
            service %s", svc->current_state, svc->host_name,
            svc->description );
```

```
write_to_logs_and_console( outbuf , NSLOG_INFO_MESSAGE, TRUE );

    //create the host and service path strings
    sprintf( host_path,"/usr/share/nagios/status/%s", svc->host_name);

sprintf( service_path,"/usr/share/nagios/status/%s/%s",
        svc->host_name, svc->description);

    //create the directory
    mkdir( host_path, S_IRWXU|S_IRGRP|S_IXGRP|S_IROTH|S_IXOTH );

    //write the file
    FILE *outfile=fopen(service_path,"w");
    fprintf(outfile,"%i:%i",svc->current_state,svc->state_type);
    fclose(outfile);
    return 0;
}
```

To compile this, you'll need access to the headers that come in the include directory of the Nagios tarball. Grab and extract the current tarball, and then run configure and make all. After this is done, you can create a new directory for your module in the root of the tarball directory. Place this file in there as something such as `fs.c` and compile it into a shared lib with the following:

```
gcc -shared fs.c -o fs.o
```

Decide on a decent location for NEB modules for your particular distro packaging, and copy the object file there. All that's left is to tell Nagios to load the module. Do this by adding the following line to your nagios.cfg:

```
broker_module=/path/to/your/modules/fs.o
```

For Nagios to load the module, you need to have compiled Nagios with Event Broker support. See Chapter 4, "Configuring Nagios," for details on doing that. If all goes well, you should see something similar to the following in your nagios.log when you start up Nagios:

```
[1156799404] Loading FS Module…
```

Now, let's look at the code starting at the top, with the include statements in Listing 10.3.

Listing 10.3 *Includes*

```
/* include the needed Event Broker header files */
#include "../include/nebmodules.h"
#include "../include/nebcallbacks.h"
#include "../include/nebstructs.h"
#include "../include/broker.h"

/* include some Nagios stuff as well */
#include "../include/config.h"
#include "../include/common.h"
#include "../include/nagios.h"
#include "../include/objects.h"
```

The first section of headers includes the data structures and functions necessary to interact with the Event Broker. These encompass functions like neb_register_callback for subscribing to interesting events and structs like nebstruct_service_status_data, which contains the notification data handed down from the Event Broker for a service status event. The second section of headers describes core Nagios functions and structs, so things such as service_struct, which define a Nagios service, and write_to_logs_and_console, which is a function for… well, writing messages to the logs and console.

If you plan to write an Event Broker module, you'll have to poke around in most of these headers, but even if you don't plan to write a module, I encourage you to take a look anyway. Nagios is one of the most nicely written C programs I've had the pleasure of digging around in. The function and variable names are self-documenting, the comments are terse, descriptive, and liberally dispersed, and the application, as a whole, is well engineered.

After the include statements is a declaration for our handler function. I declared it at this point in the program so that I wouldn't have to write about it yet, so ignore it for now. The next relevant section is in Listing 10.4.

Listing 10.4 *Some Required Tidbits*

```
/* specify Event Broker API version (required) */
NEB_API_VERSION(CURRENT_NEB_API_VERSION);

// Our module handle
void *basic_module_handle=NULL;
```

The NEB_API_VERSION macro is designed to ensure that the module is running under the version of the NEB API that it was compiled to run on. All NEB modules are required to include this line. The void pointer declaration is a globally scoped handle that will eventually refer to the memory address of our own module. It's declared here so that it can be referenced in a global context. It's important that we are able to refer to it ourselves, later, when we start registering for callbacks, such as in Listing 10.5.

Listing 10.5   *The init Function*

```
/* this function gets called when the module gets
loaded by the Event Broker*/
int nebmodule_init(int flags, char *args, nebmodule *handle) {

   basic_module_handle = handle;

write_to_logs_and_console("Loading FS Module...",
                     NSLOG_INFO_MESSAGE,TRUE);

neb_register_callback(NEBCALLBACK_SERVICE_STATUS_DATA, handle,
                     '1', (void *)&handle_service_status);

   write_to_logs_and_console("Done",NSLOG_INFO_MESSAGE,TRUE);

   return 0;
}
```

Every NEB module is required to have an entry function and an exit function. Listing 10.5 is the entry function for our module. As you can see, it returns an exit code in the form of an int and takes three arguments. The first argument, an int called flags, is meant to give you the context in which the module is being initialized. The second argument is a string pointer called args. It is possible to pass arguments to your module in the nagios.cfg file by adding them to the end of the module name in the broker_module directive. For example, our current filesystem module hard-codes its base directory as /var/lib/nagios/status, but we could have passed the directory name as an argument to the module with the following definition in the nagios.cfg:

```
broker_module=/path/to/your/modules/fs.o \
base=/usr/lib/nagios/status
```

I chose not to do this, to keep the example source code simple. But had I specified it as an argument, I could have parsed it out with the args pointer in the init function. The third argument is a pointer of type nebmodule that points to a struct, which defines our module. In short, this is a handle that uniquely identifies the memory address for our module. The nebmodule struct is defined in nebmodule.h, and, in case you're curious about what a module consists of, Listing 10.6 contains the definition.

Listing 10.6   *The nebmodule struct*

```
/* NEB module structure */
typedef struct nebmodule_struct{
    char            *filename;
    char            *args;
    char            *info[NEBMODULE_MODINFO_NUMITEMS];
    int             should_be_loaded;
    int             is_currently_loaded;
#ifdef USE_LTDL
    lt_dlhandle     module_handle;
    lt_ptr          init_func;
    lt_ptr          deinit_func;
#else
    void            *module_handle;
    void            *init_func;
    void            *deinit_func;
#endif
#ifdef HAVE_PTHREAD_H
    pthread_t       thread_id;
#endif
    struct nebmodule_struct *next;
        }nebmodule;
```

As I said, Nagios is a nifty C program to work with. There are all sorts of interesting tidbits of information we could query about our own module, and other functions will surely want to know some of this stuff about us. So the first thing we do in the init function is to cache a copy of our own memory handle with the following line:

```
basic_module_handle = handle;
```

After we've done that, we could start dereferencing information about ourselves, if we so desired. For example, if we wanted to know our thread ID, we could do something like the following:

```
pthread_t t_id=handle->thread_id
```

After we have a copy of our handle, we write some output to the console and log files to let the outside world know that we are alive and functional. This is done by the write_to_logs_and_console() function, which is defined in nagios.h. This function takes three arguments; the first is a string that points to the message. The second is a constant that defines the type of message; there are several of these, also specified in nagios.h. The one we use is for informational messages. The last argument is a Boolean that toggles console output, so if this is true, the message goes to the console, as well as the logs.

With the next line, we register with the Event Broker to receive some events:

```
neb_register_callback(NEBCALLBACK_SERVICE_STATUS_DATA, handle,
                      '42', (void *)&handle_service_status );
```

The `neb_register_callback` function is defined in nebcallbacks.h. The definition looks like this:

```
int neb_register_callback(int callback_type, void *mod_handle,
                          int priority, int (*callback_func)
                          (int,void *));
```

So `neb_register_callback` takes four arguments. The first is the constant describing what types of events we're interested in. I listed these in Table 10.1. The second is our handle, so the broker can find out what it needs to find out about us. The third is a priority number. In general, when more than one module registers for the same type of event, they are executed, in the order they are loaded, by the broker on startup. You can override this behavior by specifying a priority number.

The last argument to `neb_register_callback` is a function pointer, as described in the section "Function References and Callbacks in C." The function pointer must point to a subroutine that returns an exit code in the form of an int and accepts two arguments. The first of these is a constant specifying the event type; yes, again, one of the constants specified in Table 10.1. The second is a void pointer, which I'll get to in a moment. So, in short, this last argument to `neb_register_callback` is the function to which the Event Broker will send the event. It is the event handler.

But why would our event handler need to be passed the event type? The event handler function should be able to infer the event type by virtue of the fact that we specified which events we were interested in at the same time we defined the handler. Our event handler example was written to specifically handle the one type of event that we registered for, but this isn't necessarily a requirement. The nice thing about being passed back the event type constant is that it enables the module to register for more than one type of callback and handle each type it registers for with a single event handler function.

So what's this null pointer for? To answer that, we need to look at what the Event Broker is doing when it makes our callback. Consider the code in Listing 10.7 from broker.c in the base directory of the tarball.

Listing 10.7 *The Event Broker Sending Data*

```
/* sends program data (starts, restarts, stops, etc.)
to broker */
void broker_program_state(int type, int flags, int attr,
                          struct timeval *timestamp){
   nebstruct_process_data ds;

   if(!(event_broker_options & BROKER_PROGRAM_STATE))
      return;

   /* fill struct with relevant data */
   ds.type=type;
   ds.flags=flags;
   ds.attr=attr;
   ds.timestamp=get_broker_timestamp(timestamp);

   /* make callbacks */
   neb_make_callbacks(NEBCALLBACK_PROCESS_DATA,(void *)&ds);

   return;
      }


/* send timed event data to broker */
void broker_timed_event(int type, int flags, int attr,
                        timed_event *event,
                        struct timeval *timestamp){
   nebstruct_timed_event_data ds;

   if(!(event_broker_options & BROKER_TIMED_EVENTS))
      return;

   if(event==NULL)
      return;

   /* fill struct with relevant data */
   ds.type=type;
   ds.flags=flags;
   ds.attr=attr;
   ds.timestamp=get_broker_timestamp(timestamp);

   ds.event_type=event->event_type;
   ds.recurring=event->recurring;
   ds.run_time=event->run_time;
   ds.event_data=event->event_data;

   /* make callbacks */
   neb_make_callbacks(NEBCALLBACK_TIMED_EVENT_DATA,
                      (void *)&ds);

   return;
      }
```

There's a pattern here. We see two functions, which represent two types of events (again, defined by the constants in Table 10.1) being sent out by the broker. In each case, the broker first populates a struct called ds with data relevant to the type of event it is about to send, and then, after the ds struct is populated, it uses the neb_make_callbacks function to send the event-type constant and a pointer to the ds struct. The broker does this same thing for every type of event in Table 10.1, so the null pointer our event handler function receives is the ds struct that the broker populates. This data is specific to each type of event; for example, when the broker makes a callback to the modules interested in timed events, the ds struct is of type nebstruct_timed_event_data.

After our init function in Listing 10.5 registers for events of type SERVICE_STATUS_DATA, it writes the word "done" to the logs and exits with a "0", the universal sign that everything's "a-okay." The deinit function that follows init, in our example in Listing 10.2, is also required, but doesn't bear much of an explanation. It receives some constants that specify the reason the module is being unloaded and provides you an opportunity to do some cleaning up before your module goes bye-bye.

Now we're ready to take a look at the event handler function in Listing 10.8. Most of the guts of our program reside there.

Listing 10.8   *Our Event Handler Function*

```
int handle_service_status(int neb_event_type, nebstruct_service_
➥status_data *ds){

    //get a handle to the service struct
    service *svc = ds->object_ptr;

    //create some name buffers for various output
    char outbuf[100];
    char host_path[100];
    char service_path[100];

    //create a string with some logging info
    sprintf( outbuf,"Caught status code: '%i' from host '%s' for
            service %s", svc->current_state, svc->host_name,
            svc->description );

//write it out to the logs
write_to_logs_and_console( outbuf , NSLOG_INFO_MESSAGE, TRUE );

    //create the host and service path strings
    sprintf( host_path,"/usr/share/nagios/status/%s",
            svc->host_name );
    sprintf( service_path,"/usr/share/nagios/status/%s/%s", svc->host_
➥name, svc->description);
```

```
//create the directory
mkdir( host_path, S_IRWXU|S_IRGRP|S_IXGRP|S_IROTH|S_IXOTH );

//write the file
FILE *outfile=fopen(service_path,"w");
fprintf(outfile,"%i:%i",svc->current_state,svc->state_type);
fclose(outfile);
return 0;
}
```

As I said a bit earlier, our event handler will be passed two arguments: the event type constant and a reference to the ds struct containing information about our specific type of argument. But what will the ds struct contain for events of type SERVICE_STATUS? Let's take a look at the relevant code snippet in Listing 10.9, from the broker.c.

Listing 10.9   *The Broker's 'make_callback' code for SERVICE_STATUS_DATA*

```
/* sends service status updates to broker */
void broker_service_status(int type, int flags, int attr,
                           service *svc,
                           struct timeval *timestamp){
  nebstruct_service_status_data ds;

  if(!(event_broker_options & BROKER_STATUS_DATA))
    return;

  /* fill struct with relevant data */
  ds.type=type;
  ds.flags=flags;
  ds.attr=attr;
  ds.timestamp=get_broker_timestamp(timestamp);

  ds.object_ptr=(void *)svc;

  /* make callbacks */
  neb_make_callbacks(NEBCALLBACK_SERVICE_STATUS_DATA,
                     (void *)&ds);

  return;
      }
```

The first thing to take note of is the line:

```
nebstruct_service_status_data ds;
```

This tells us that the ds struct for our event will be of type nebstruct_service_
status_data. The struct appears to have five elements: type, flags, attr, a timestamp, and a
void pointer to svc, which is a struct describing a Nagios service. Let's check out nebstructs.h
(see Listing 10.10) for a description of the struct.

Listing 10.10   *The nebstruct_service_status_data struct*

```
/* service status structure */
typedef struct nebstruct_service_status_struct{
    int             type;
    int             flags;
    int             attr;
    struct timeval  timestamp;

    void            *object_ptr;
        }nebstruct_service_status_data;
```

So we have three ints, a timeval struct, and a pointer to the service itself. The service
pointer sounds interesting. Considering all the cool stuff our module struct contained, the
service struct must really have a bunch of goodies. This leads us to Listing 10.11, which is the
service_struct definition from objects.h.

Listing 10.11   *The service_struct Def from nagios.h*

```
/* SERVICE structure */
typedef struct service_struct{
   char    *host_name;
   char    *description;
        char    *service_check_command;
   char    *event_handler;
   int    check_interval;
   int     retry_interval;
   int    max_attempts;
   int     parallelize;
   contactgroupsmember *contact_groups;
   int    notification_interval;
   int     notify_on_unknown;
   int    notify_on_warning;
   int    notify_on_critical;
   int    notify_on_recovery;
   int     notify_on_flapping;
   int     stalk_on_ok;
   int     stalk_on_warning;
   int     stalk_on_unknown;
   int     stalk_on_critical;
   int     is_volatile;
```

```
    char    *notification_period;
    char    *check_period;
    int      flap_detection_enabled;
    double  low_flap_threshold;
    double  high_flap_threshold;
    int      process_performance_data;
    int      check_freshness;
    int      freshness_threshold;
    int      accept_passive_service_checks;
    int      event_handler_enabled;
    int    checks_enabled;
    int      retain_status_information;
    int      retain_nonstatus_information;
    int      notifications_enabled;
    int      obsess_over_service;
    int      failure_prediction_enabled;
    char    *failure_prediction_options;
#ifdef NSCORE
    int      problem_has_been_acknowledged;
    int      acknowledgement_type;
    int      host_problem_at_last_check;
#ifdef REMOVED_041403
    int      no_recovery_notification;
#endif
    int      check_type;
    int    current_state;
    int    last_state;
    int    last_hard_state;
    char    *plugin_output;
    char    *perf_data;
    int      state_type;
    time_t   next_check;
    int      should_be_scheduled;
    time_t   last_check;
    int    current_attempt;
    time_t   last_notification;
    time_t   next_notification;
    int      no_more_notifications;
    int      check_flapping_recovery_notification;
    time_t   last_state_change;
    time_t   last_hard_state_change;
    time_t   last_time_ok;
    time_t   last_time_warning;
    time_t   last_time_unknown;
    time_t   last_time_critical;
    int      has_been_checked;
    int      is_being_freshened;
    int      notified_on_unknown;
    int      notified_on_warning;
    int      notified_on_critical;
    int      current_notification_number;
    double  latency;
    double  execution_time;
```

```
    int     is_executing;
    int     check_options;
    int     scheduled_downtime_depth;
    int     pending_flex_downtime;
    int     state_history[MAX_STATE_HISTORY_ENTRIES];    /* flap
➥detection */
    int     state_history_index;
    int     is_flapping;
    unsigned long flapping_comment_id;
    double  percent_state_change;
    unsigned long modified_attributes;
#endif
    struct service_struct *next;
    struct service_struct *nexthash;
    } service;
```

Wow, jackpot! The service `struct` has everything we could hope to know about a service and then some. And the nifty Event Broker handed us our own pointer, straight to the service to which our event is currently in reference. So, getting back to Listing 10.8, the first thing our event handler function does is to grab a type specific handle to the service pointer for convenient dereferencing:

```
service *svc = ds->object_ptr;
```

Then, because we'll be mixing and matching output from various sources, we create a few output buffers. I can't wait to dereference something from my cool new service handle, so the next line

```
sprintf( outbuf,"Caught status code: '%i' from host '%s' for service
%s", svc->current_state, svc->host_name, svc->description );
```

builds a string suitable to output to the logs. After we write that out, we create two more strings, which we'll use to create our fs interface. The first dereferences the hostname to which the service refers. We'll use this to create the directory for our service. The second string dereferences the service description, which we use for the filename:

```
sprintf( host_path, "/usr/share/nagios/status/%s", svc->host_name);
sprintf( service_path, "/usr/share/nagios/status/%s/%s",
svc->host_name, svc->description);
```

After this is done, we can create our directory, write our file, and we're done.

```
//create the directory
mkdir( host_path, S_IRWXU|S_IRGRP|S_IXGRP|S_IROTH|S_IXOTH );

//write the file
FILE *outfile=fopen(service_path,"w");
fprintf(outfile,"%i:%i",svc->current_state,svc->state_type);
fclose(outfile);
```

Astute readers will notice that the actual contents of the file are two numbers separated by a colon. With such a great service struct, I couldn't resist pulling out an extra tidbit of information. This is the `state_type`, which is an int that specifies whether the service is in a hard state (0) or a soft state (1). So, if ping on host15 was in a hard critical state, the contents of the file /var/lib/nagios/status/host15/ping would be '2:0.'

Although it compiles and works okay, this module has some pretty big problems. It doesn't do things such as checking whether the directory creations and file writes succeed, and it could be far more efficient about making system calls. For example, we could dereference `last_state` and compare it with current_state, to determine if it was worth opening the file. In fact, subscribing to state change events would probably be more efficient, but in the interest of keeping the example as straightforward as possible, a lot of potential functionality was omitted.

DNX, a Real-World Example

In Chapter 7, "Scaling Nagios," I introduce DNX, an NEB module that makes it possible to scale Nagios by distributing its service checks to worker nodes running on remote systems. In this section, I'd like to take a closer look at how this is accomplished by using what we now know to examine a few pieces of the DNX broker module. DNX is a too complex a piece of software—compared to the filesystem module I've been talking about so far—to provide a complete code listing, and the details of the client and daemon interaction (which happen outside of Nagios's address space) are outside the scope of what I want to talk about here. However, I can give you a good idea of where to start inspecting it for yourself, and also point out a few interesting things the DNX authors did, such as preempting Nagios's built-in service-check logic.

I start by downloading the current version of DNX (0.20.1 at this moment) from http://sourceforge.net/projects/dnx/files/dnx-0.20.1.tar.gz/download. After unpacking it, the first thing I look for is the nebmodule_init function, because I want to see what the module does when it starts up (and also I want to know what file in the tarball implements the neb module). This is easily accomplished with grep:

```
grep -rl 'nebmodule_init(' .
```

This command returns the name of the file containing the module code, which is ./plugin/ dnxNebMain.c. Examining the init function, we see a few interesting things. First is that they've implemented their own logging functions. So instead of using the Nagios built-in `write_to_logs_and_console` function, like the filesystem module, they will use their own function `dnxLog`. This function, implemented in common/dnxlogging.c, provides a common logging interface for the pieces of DNX that are inside the Nagios address space like the NEB module, as well as the pieces such as the daemon, which are external.

The next thing I notice is the `neb_register_callback` function. It subscribes to a rather unexpected callback type: `NEBCALLBACK_PROCESS_DATA`, which returns events related to the Nagios daemon process itself, such as "Nagios is starting up now," or "Nagios is shutting down now." Above the `register_callback` function is a comment that informs us of the following:

```
Subscribe to PROCESS_DATA call-backs in order to defer initialization
until after Nagios validates its configuration and environment
```

This is interesting, because DNX has an external daemon process that is initialized by the event broker module. The problem is, when Nagios initializes the module, it hasn't finished reading in all its configuration data yet, so if DNX were to bring up its daemon immediately, the DNX daemon wouldn't have an accurate depiction of which hosts and services were configured. For this reason, the DNX module subscribes to the `PROCESS_DATA` event type and waits until it sees an event from the Nagios daemon that Nagios is finished reading in its entire configuration. When DNX sees that Nagios is finished, it will bring up its own daemon and continue with its own processing.

This register_callback function gives `ehProcessData` as the handler function, so the next thing I do is take a look at that function, because I want to see how they've implemented this deferred initialization. The code is simple enough:

```
// look for process event loop start event
if (procdata->type == NEBTYPE_PROCESS_EVENTLOOPSTART)
{
    dnxDebug(2, "Startup handler received PROCESS_EVENTLOOPSTART
➥event.");
```

```
    // execute sync script, if defined
    if (cfg.syncScript)
    {
        dnxLog("Startup handler executing plugin sync script: %s.",
➥cfg.syncScript);

        // NB: This halts Nagios execution until the script exits...
        launchScript(cfg.syncScript);
    }

    // if plugin init fails, do plugin shutdown
    if (dnxPluginInit() != 0)
        dnxPluginDeInit();
}
```

The Nagios daemon signals that it is ready using the EVENTLOOPSTART constant, and DNX runs the dnxPluginInit function in response. It can be safely assumed that DNX will perform additional event registrations inside this PluginInit function, but why assume when we can just go look (see Listing 10.12).

Listing 10.12 *dnxPluginInit() Function*

```
static int dnxPluginInit(void)
{
    int ret;
    // start dnx server child process
    if ((ret = execServerProcess()) != 0)
    {
        dnxLog("Error starting DNX server process: %s.",
➥dnxErrorString(ret));
        return ret;
    }
    dnxLog("DNX Server process started.");
    // start results listener
    if ((ret = pthread_create(&rlthread, 0, dnxResultsListener, (void
➥*)0)) != 0)
        return ret;
    dnxLog("Results listener thread started.");
#if CURRENT_NEB_API_VERSION == 3 && defined(DIRECT_POST)
    // register for timed event to piggy-back on reaper thread
    neb_register_callback(NEBCALLBACK_TIMED_EVENT_DATA, myHandle, 0,
➥ehTimedEvent);
    dnxLog("Registered for TIMEDEVENT_EXECUTE event.");
#endif
    // registration for this event starts everything rolling
    neb_register_callback(NEBCALLBACK_SERVICE_CHECK_DATA, myHandle, 0,
➥ehSvcCheck);
```

```
    dnxLog("Registered for SERVICE_CHECK_DATA event.");
    dnxLog("DNX Nagios Plugin initialization completed.");
    return OK;  // Nagios OK value
}
```

As you can see, my assumption panned out; DNX registers for two additional event types. The first is `TIMED_EVENT_DATA`, which it uses to synchronize the results it receives from its worker nodes to Nagios's naturally occurring check results reaper process. Note the use of the C preprocessor to test the current version of the event broker API. This is a handy technique for ensuring portability with different Nagios versions. The second event type DNX registers for is `SERVICE_CHECK_DATA`, which it uses to detect and preempt Nagios service checks.

The last thing I'd like to show you is how that preemption is accomplished. Taking a look at `ehSvcCheck`, which is the event handler function from the register function that subscribes to `SERVICE_CHECK_DATA` in Listing 10.12, we find that DNX performs various tests to make sure that the service check in question matches the type of service check it's interested in distributing, and if it is, the following code is executed:

```
// try to post this job to the dnx server process
if ((ret = dnxPostNewJob(serial, svcdata)) != DNX_OK)
{
   dnxDebug(1, "Post failed: %s. Service check [%lu] will execute
➥locally: %s.",
        dnxErrorString(ret), serial, svcdata->command_line);
   return OK;
}
return NEBERROR_CALLBACKOVERRIDE;   // tell nagios we want it
```

The `dnxPostNewJob` function is passed a pointer to the svcdata struct, along with a unique serial number that DNX computed earlier in the `ehSvcCheck` function (a simple iterator on a counter variable). If `dnxPostNewJob` fails, this function returns OK and Nagios will pick up where it left off. If, however, it succeeds, the function returns `NEBERROR_CALLBACKOVERRIDE`, which is a special exit code notifying Nagios that it should abandon its normal service check logic and assume that DNX has taken care of it.

Wrap Up

My goal, in this chapter, was to get you excited about the Event Broker interface and what it is capable of. In that, I hope that I succeeded, because the Nagios community needs creative folks like you to contribute interesting modules. If you have an idea for a useful module, I hope this chapter gave you a head start, and I can't wait to use it when you're done.

End Notes

[1] This is by design. In Ethan's words: "There are a whole number of things that I would like to see Event Broker able to do. Essentially, I would like to allow Event Broker modules to override most of the internal logic in the daemon when it comes to host/service checks, notifications, flap detection, logging, executing external commands, et cetera. This will allow people to do a number of neat things that would otherwise require extensive rewriting of the Nagios daemon."

[2] Actually, we could solve this one more easily with the global event handler, but it makes a good NEB example.

INDEX

Symbols

.1.3.6.1 prefix, 135
* (asterisk), 67
{} (curly braces), 64
$ (dollar signs), 96
. (dot), 116

A

abnormal utilization, 169
acknowledgement, notification, 43
action_url, 198-199
active_checks_enabled, 150
address directive, 75
Afterglow, 218
alarms, false alarms, 19
Alert summary, 47
Apache, configuration, 83-85
applications versus ports, watching, 20-22
architecture
    Event Broker, 239-241
    Nagios XI, 225
AREA, RRDTool, 181
argument passing, command definitions (check_load), 128
asterisk (*), 67
authorized_for_all_host_commands, 68
authorized_for_all_hosts, 68

authorized_for_all_service_commands, 68
authorized_for_all_services, 68
authorized_for_configuration_information, 68
authorized_for_system_information, 68
authorizied_for_system_commands, 68
autodiscovery, 91-92
    Check_MK, 91
    Nagios XI, 92
automated configuration for passive checks, 233
AVERAGE, RRDTool, 180
averageSeries, 209
awk, 196

B

backups, Nagios XI, 234
bandwidth, monitoring systems, 13-14
bar charts, data visualization, 210
baselines, 19
benefits of Nagios XI
    advanced reporting and advanced visualization, 228-230
    integrated plug-ins and configuration wizards, 230-233
    integrated time series data, 227-228
    interface, 226-227
    modularized components, 228
    operational improvements, 234

FREE
Online Edition

Your purchase of *Nagios* includes access to a free online edition for 45 days through the **Safari Books Online** subscription service. Nearly every Prentice Hall book is available online through **Safari Books Online**, along with thousands of books and videos from publishers such as Addison-Wesley Professional, Cisco Press, Exam Cram, IBM Press, O'Reilly Media, Que, Sams, and VMware Press.

Safari Books Online is a digital library providing searchable, on-demand access to thousands of technology, digital media, and professional development books and videos from leading publishers. With one monthly or yearly subscription price, you get unlimited access to learning tools and information on topics including mobile app and software development, tips and tricks on using your favorite gadgets, networking, project management, graphic design, and much more.

Activate your FREE Online Edition at
informit.com/safarifree

STEP 1: Enter the coupon code: IOQKGWH.

STEP 2: New Safari users, complete the brief registration form. Safari subscribers, just log in.

If you have difficulty registering on Safari or accessing the online edition, please e-mail customer-service@safaribooksonline.com